HARALD SEIZ

THINK
BIG

HARALD SEIZ

THINK
BIG

How to conquer the World
with a great idea

FBV

The Deutsche Nationalbibliothek
Lists this publication in the Deutsche Nationalbibliographie; detailed bibliographic information is
available online at http://dnb.d-nb.de abrufbar.

1st Edition 2019

© 2019 by FinanzBuch Verlag,
an imprint of the Münchner Verlagsgruppe GmbH,
Nymphenburger Straße 86
D-80636 München
Tel.: 089 651285-0
Fax: 089 652096

Translation: Shelley Steinhorst
Proof Reading: Hella Neukötter
Cover Design: Karatbars International GmbH, Manuela Amode
Cover Illustration: CCO Creative Commons
Type Setting: ePubMATIC LLC
Printing House: GGP Media GmbH, Pößneck
Printed in Germany

ISBN Print 978-3-95972-150-9
ISBN E-Book (PDF) 978-3-96092-272-8
ISBN E-Book (EPUB, Mobi) 978-3-96092-273-5

Further informations are available at

www.finanzbuchverlag.de

Please note our other publishers at www.m-vg.de

CONTENT

FOREWORD

Everyone starts out small. The global corporations of today were once local family businesses or simply a thought in the mind of an innovative businessman. But that's where everything starts: in the mind. Add creativity and some business savvy and no goal is too far out or too far-fetched—success is not utopia. This is what I will show you in this book.

I want to show you how to succeed in taking your great idea from the realm of the mind into a life that's real. Of course, that is not always easy. There will be obstacles along the way. As multifaceted as the world is today, so are the problems that you can expect to encounter on your way to the top. Financial deficits, personal loss, criticism and accusations from friends and business partners. There is nothing in the world that hasn't already taken place along the path towards success as an entrepreneur.

But these obstacles are not insurmountable hurdles. There are always solutions. You just have to recognize them. You can view the challenges as problems or you can see them as opportunities, as a chance to develop yourself and your project. This is the true spirit of entrepreneurship.

How this can work for you and your situation is crystallized in the following six chapters. I have had to deal with setbacks and disappointments, myself—both professionally and personally. I took on this fight and made the best of my life. What you are about to read in this book is based on my own life experience. Experiences that I am happy to pass on to you to help you master your own personal way to success. Because I am convinced that you can achieve greatness with a single, great idea.

From the initial idea, to setting up your own company, and to dealing with the success, this book is a guide to achieving your career independence. A career that can sometimes be a balancing act—between doubt and success, between motivation and stagnation. But I'm not going to focus on abstract concepts. I'll leave the hypothesizing to others. I'm a practical man and that's how I've written this book. In touch with reality and based on real life. This is an invitation for you to take everything you need from my examples and personal anecdotes and create a world-wide imperium from your own great idea. In this spirit, I wish you an exciting and productive read.

I.
"YOU MUST BURN FOR IT TO BE ABLE TO IGNITE THE FIRE IN OTHERS."

From visionary to millionaire—it sounds unbelievable. But vision is at the beginning of every successful career. A dream and a plan of how to create your life. Like a captain guiding his ship to its port through both quiet and volatile waters. With his sights set to the stars in the sky and his intuition as a compass. When I first gazed into the future I saw only an endless expanse. But even in this great vastness, the first benchmarks came into focus that would later become the cornerstones of my career. My "stars" were as different as they were authentic. Dr. Martin Luther King inspired me with his famous statement, "I have a dream," with its unbridled rhetorical power and his revolutionary drive for a better world for his people. He dared to talk openly about things that others hadn't allowed themselves to think. He put himself at the vanguard of a movement and stayed true to his dream up to the bitter end. Full of conviction, he fought for what he believed. He set tremendous societal forces in motion, becoming immortalized in the hearts of many. Like virtually no other, he is an enduring example of how far one can go when one is in it with heart and soul.

He put his mark on a country that I enjoy touring, today. In spite of all the criticism that you hear in the current Trump atmosphere, I am a big fan of the American way. The country lives the spirit of freedom and independence. I have always enjoyed being in the USA and exploring its far corners and meeting new people. My second big idol is also an American. Elvis Presley, the "King of Rock'n'Roll," with his unbelievably incredible voice. He showed me how dreams can come true with passion and a powerful presence. The conditions that he grew up in were anything but easy, and his childhood not anywhere close to what you'd call carefree. His twin brother died during birth and his parents worked on a farm and in a factory with very modest means. And though there was not much money for recreational activities, the family made the best of the little they had for young Elvis. He grew up in the small town of Tupelo in Mississippi. The technical advances of the time had not yet reached this area of the world. Almost no one in Tupelo owned a radio, let alone a television.

Instead the Presley house was the epicenter of entertainment in the neighborhood. As creative as they were, there was sure to be a good time at the Presley's with their gospel choir. With little Elvis in the middle of it all, his sonorous voice soon became the main attraction in the little town, and later in all of the United States.

Like Elvis, I also grew up in modest circumstances. Born in Stuttgart, Germany, in 1963, I had to bid farewell to my father at the age of three, when he decided to return to his hometown in Greece. Although I had my mother and grandmother at my side, I had to learn to stand on my own two feet at an early age. This difficult family situation meant it was not an easy time for me. In these days of patchwork families and open marital partnerships, it's hard to believe how different family life was in a West German province in the 1960's. Patterned wallpaper, furniture that one could only find in a retro shop, today, and rigid definitions of how a "good" family behaved—this all characterized the time. In the 1960's, a single mother was not at all common practice. On the contrary, they were regarded quite critically. Their lifestyle was constantly standing trial. Over and over again I sensed the skepticism that my family was greeted with. As little as I was, I could sense the looks of the neighbors and noticed their questions: Where is the husband? Why did he leave his wife? How can a mother take care of her son alone and also go to work without neglecting him? The West Germany of the 1960's, that I got to know, was characterized by a conservative family image in which the father was supposed to take on the entire responsibility for the family. Only, in this case, my father had run for the hills. If I learned anything from that time, it was that one should never allow oneself to depend on the opinions of others. Regardless of how stifling or oppressive the majority opinion might be, or how massive the social peer pressures are: you have to stay true to yourself. Who would have believed that the little boy from the small town Weil, the child of an immigrant without a father figure, would someday be running a company worth millions? No one! And yet I managed to accomplish this against all odds. Because I went my own way—I ignored the advice and the reproach of others. And this path lead me to become the successful entrepreneur that I am, today.

There were plenty of obstacles and hurdles along the way—personally and professionally, from a very early age. I had to learn quickly to be completely self-reliant. The classical family model with the father as the "provider" and the mother as "carer," as was usual in those days, didn't exist at our home. My mother and my grandmother worked in a nearby factory to make ends meet. They saved every penny, and didn't waste a dime. The work on the assembly line didn't allow for much time to look after me, so they sent me to a daycare center in a small, sedate little town nearby. Every weekday we took the first train at 5:30am. While most other kids were still slumbering, deep in their dreams, I was all ready to step out the door. And I was full of energy. I wanted to go out and discover the world and gather more experience. This was a curiosity that still leads me to new shores, today.

This curiosity would find a new source of nourishment when I started school at the age of six. I expected a lot from school, wanted to learn new things. But it was an uphill battle. It all started when I realized after a few days of school that I couldn't read the letters of the alphabet correctly on the blackboard. They blurred into an illegible white mass against a green backdrop. From then on and after a short visit to an optician it became clear: I needed glasses. But I didn't get the usual glasses. No, I had to wear huge, chunky, black, thick-rimmed glasses. They made me shrink in comparison. And before I knew it, I was the class "four-eyes." It started when I was on the way to school. The other kids would start taunting "four-eyes, four-eyes!" No one wanted to play with me. Once, when I asked a kid in my class if he wanted to play, he said, "First learn to look where you're going." The taunting carried on in the playground and in the classroom. At first, the teacher tried to defend me from the tormenting. But as soon as she turned her back on the class, or wasn't nearby, they carried on with the "four-eyes, four-eyes!" Kids can be so cruel.

My situation in school was not that enviable. One might say: "At least the boy could gather strength from within the bosom of the family." And to a certain extent that was true. At the same time, however, there were problems waiting for me at home that I had to come to terms with very early on.

Even as a young boy I noticed that there was something funny about my mother and grandmother. One day they would be in ecstatic high-spirits, without any apparent reason, at least that I could see. The next day they would be worn out, lying in bed, tired and "hungover." On those days, it was impossible to even get them to get up. At first, I could think of no explanation for their mood swings. I was just too young for these things. But later I understood that it was because of the alcohol. But why did they start to drink excessively, in the first place?

Both my mother and my grandmother had their burdens to bear. Experiences that they could never really process and come to terms with. During the Second World War my grandmother had had an affair. Her husband left her at the end of the war. My mother had also been left—by my father, from whom I have not heard a word, not even today. These experiences left their marks. They had to suppress what had happened to them. And they found consolation in alcohol. I remember going to get beer for them, and that they drank without reserve twice a month.

Once I came home to find my mother on the floor. She lay there unmoving, murmuring unintelligibly to herself. Blood was running down her head. I felt the panic rise. What had happened? I began to feel mortal fear. Fear that my mother would die. I called for help but my voice died away, unheard. I continued to call out: "Help, help, my mother...!" I began to shout and just before I thought my voice would choke from hoarseness, I heard a knock at the door. I ran to the door and opened it. Our neighbor was standing there. I looked up into his shocked face: "What's going on?" I showed him my mother. He reacted immediately and it didn't take long before the ambulance sirens could be heard. My mother was brought to the hospital. I worried, prayed, hoped that everything would turn out okay. And I was lucky. My mother survived. But I had been scarred by this event. I found it hard to concentrate at school. I was always worried that something could be happening at home and my mother was having problems. This uncertainty wore me down, diverted my attention. And as soon as I managed to turn my focus back to the school material, I would again become aware of the taunting, the comments and insults of my school mates. It was horrible!

But as bad as it was, inside myself as well—I rose to the challenge and learned to concentrate on the positive things in life. I recognized the sheer infiniteness of opportunities that life has in store for us. Although, as a young boy I did not own much from a materialistic point of view, I cherished a real treasure from within. With my imagination and a great passion, I felt a huge desire to change the world, one step at a time, and not give up until I had fulfilled my dreams, like my idols in the USA. You don't need a huge fortune for that. All you need to do is believe in yourself and be courageous even in the face of setbacks. The path from the bottom to the top is not a utopian idea. It is also not a one-way street with no oncoming traffic. Sometimes it felt like having to scale a mountainside the size of the Alps, sometimes a dense jungle of prohibitive rules and regulations. But those times pass, and they are a relief.

When my mother met the man who later became my stepfather, more stability returned to our lives. Like me, he also had

a migrant background. He fled his home in the Czech Repub-
lic to come to West Germany. Like so many others, he was looking for a better livelihood and more freedom. He didn't move to
Germany alone. He came with his parents and three brothers.
All were extremely hardworking, keeping their own businesses
going. My stepfather was a painter and was always busy with work.
He profited from the economic growth in Germany at the time.
Baden-Württemberg was a boom federal state then, with building and renovations going on all over the place. While it didn't
bestow a luxurious income on us, it was a secure one, nevertheless. In keeping with the new family circumstances, we also moved
into a new place. My stepfather bought a new house in Gerlingen
together with his brother, not far from our old house. Every Sun-
day we went out on excursions together to check out the surroundings. We explored the region, enjoying the nature together. The
wide fields, the beautiful green treetops, the light along the foot-
paths—those are all happy memories from my childhood. How
often such splendor and beauty is contained in such seemingly
simple things! I almost couldn't believe it.

In school things started to go better as well. I was still being
tormented, my glasses still looked like thick window panes which
made my appearance suffer. But things finally began to get better,
because I found a very good friend by the name of Wolfgang. And
that made a lot of things easier. He was the only one in my class
who didn't care how I looked. I still remember how he came to me
during recess, without reserve or aversion. We would exchange a
few words, start to play and with each day that passed together,
we noticed that we were becoming really good friends. The mean
laughter of the others didn't interest me anymore. I had a friend
at my side. Someone who accepted me as I was. Someone who
was really important to me. The others were only trying to make
themselves seem more important by going after the "supposed"
little guy. But I was no longer alone. As important as it is to find
strength from within yourself to go your own way: no man is an
island. You need companions, supporters, business partners to
work with together to achieve your goals. I came to this realization

in grade school and still carry it with me today and have done throughout my entire career.

Wolfgang believed in me and my abilities. And I also began to recognize some of my talents. At the age of 15, and with a huge portion of self-confidence, I impersonated Elvis at a show at my favorite disco. Wolfgang was the one who gave me this chance. He had thrown my name into a raffle draw. And when my name was pulled out, there was no holding me back. I wanted to show everyone what I was made of. I let it all hang out, wild and irreverent. I was determined to express my newly won self-confidence, replacing the fear and doubt that had dictated my actions in the past, with bravery and optimism. Looking back, this experience opened doors to unexpected possibilities for me. An explosion that unleashed powers in me that I had previously only guessed at. Their initial surprise gone, my friends cheered me on ecstatically at the front of the stage. This feeling created a flood of inspiration for me. By doing something that I completely believed in, I won my friend's enthusiasm. I was perhaps not the most talented singer. I was miles away from being any serious competition for the "King of Rock'n'Roll." But I had given my all in that performance. Without any doubt and without false modesty. Considering all the energy we put into thinking about what other people think, sometimes the best way to get ahead is to just get out there and do it. Sure, my friends could have booed me from the stage and woke me up to the hard realities in life, if I hadn't been so convincing. But not even trying, is not an option. If you're not living your life because of too many "coulds," "woulds" and "should haves," you're missing out on the best opportunities life has to offer. You'll be left to look back at how other people went their way through life. But just looking on in the back row of the audience will not help you reach your goals in the long run. The "American Dream" is alive and well in anyone who recognizes their own potential.

With the right conviction, you can convince others—that became clear to me on that fateful evening. The best thing was that not only I had fun, the others were also in a good mood after this surprise performance: a true win-win situation.

My success throughout this one evening would become my life's motto. Doing good for the benefit of others would be my life's goal. Making my own life better, but above all that of my customers and fellow human beings, with the help of a good idea: that was my dream.

I had accepted that I had friends who had a much easier time in their families. Who probably received a lot more support. Who had been granted an easier time in finding their way in life. But that wasn't going to make me relinquish my chance of taking control and living according to my convictions. On stage I sensed that people don't ask where you come from. What they're interested in is who you are and whether you believe in yourself.

For a moment I forgot the problems in school and in the family. I was only aware of myself and my voice. When the first sounds came out of the speakers and I belted out the first notes, I felt free. Freed from the dead weight laying heavily on my shoulders. Full of strength and energy. I had only one thought: Now is the time! Now you can give it all and show what you're made of.

We all benefit from a gift that we carry inside of ourselves that we need to appreciate. A factory of thoughts and ideas along with a brain. Using this gift for the benefit of mankind is my most important maxim. But wouldn't it have been easier to simply wipe away these differences between me and my friends with a swipe of the hand, as if they had never existed? No, my family had made an impression on me and I have learned a lot from them. Whether voluntarily or involuntarily, because of my mother's alcoholism or the strictness of my stepfather—all of these experiences formed me into who I am today. I can still remember exactly the strict regime of my stepfather, as if I had been under his wings up until yesterday. At 8:00pm on the dot I had to be in bed. I was only allowed a measly hour more, on the weekends. I was not allowed to watch the most exciting of movies to their end. Of course, I complained, cried and screamed from anger and disappointment. But no matter what kind of fuss I made, my stepfather remained true to his word. That was hard on me, but eventually something precious developed from this strict upbringing. My daily life took on a structure that gave me the security that I hadn't had before.

Because, in contrast to other young people in my generation, I developed an intense desire for security. When you are on your own, you notice more quickly than others when things are not going well. There is no one who takes responsibility for you or can take on the blame for you. No one who jumps in to take your side. Nothing makes this point more clearly than an old saying that I learned when I was young: "The bed you make is the one you've got to lie in." As painful as that sometimes is, it is a very educational experience. An experience that cannot be replaced and is so powerful that it can completely change the direction of your life. This change was my need for security. I wanted to be armored against the worst case scenario, to build a secure bomb shelter when it came to the crunch.

And it didn't take long before it came to the crunch, again. For my little sister who arrived when I was twelve and with whom I've had a fantastic and special relationship since that day. I had just come home from school, one day, and heard my sister coughing. Not once, not twice, but continuously in a rumbling tone. I knew right away that something was wrong. She hadn't swallowed something and she didn't have a simple cold. My mother, probably under the influence of alcohol, stood next to her and asked my one-year-old sister: "What's with all the coughing?" Was she really expecting an answer? In that moment I recognized that I had to take over the responsibility. I was the only one at the time who had a clear head. I took the initiative and immediately called my stepfather. "You have to come here right now. Something's wrong with Andrea."

Luckily, my stepfather didn't waste any time in coming. We took my sister and drove straightaway to the hospital. On the way there, I sensed how it became harder and harder for her to breathe. I held her in my arms and looked deeply into her big eyes. A look so honest, sweet and innocent. This little person couldn't be allowed to die! Her breaths came in ever longer intervals, and kept getting weaker. I could sense that she was fighting for her life and tears came to my eyes. One tear after the other fell from my eyes onto the cheeks of her sweet, little face. With the boldness of desperation I screamed at my stepfather: "Drive faster!" I believe, I never again had so much fear of losing someone as in that moment. This tiny, innocent person did not deserve to die so young. And luckily everything turned out much differently. When we arrived at the hospital, the doctors met us at admissions. Everything happened so quickly and yet it felt as though we waited an eternity. The seconds passed like they were made of lead until finally the long-awaited relief came with the diagnosis. I felt my pulse racing and my head was pounding with one worry after the other. Would she survive? Did we act quickly enough? Would I ever be able to look into her eyes again, see her laugh once more, play with her again? And then finally the good news came. As the door to the examination room opened and the doctor came out with a smile on his face, all my fears and worries disappeared. We got the all-clear, but without my and our efforts, it would have been too late.

Although, this event might give one the impression that my stepfather and I were a great team, quite the opposite was true. Of course, in extreme life or death situations we could depend on each other. But daily life was something very different, altogether. As soon as I had my high school diploma under my belt, my family urged me to get a job and hire on somewhere as an apprentice. I was uncertain about what I wanted to do and finally gave in to their pressuring. I began my time as an apprentice for a bag manufacturer. We produced high-quality goods for distinctive customer tastes. They were beautiful products, without question, a perfect fit to the style of the times. But I noticed at once that this was not the right work for me. Still, I tried to master the work as best I could. My supervisor at the time, however, did not value my efforts. He didn't realize at all how much I had invested in the vocational training. Instead, he nagged away at me, dictated never-ending rules and regulations and always found something to complain about. I felt torn. Should I continue? Should I give up? What would become of me? I felt that I could do more than simply do work by the book. I needed to emancipate myself. A few days later, I again had to swallow some harsh words when I shared my plans with my boss: "I'm going now."—"You're not going anywhere." I had expected that kind of answer and came back even more decisively: "I'm going, and for good." My family were horrified by my decision. They all began to worry, and asked me and themselves what would become of me. But I knew I had to move on. I didn't have the manual dexterity for that kind of work and the monotony of the continually repeated manufacturing processes made me apathetic. I couldn't stay there any longer, so I moved on. I went through another two years of vocational training before I finally found my way into the financial industry. My second apprenticeship, a baker's training program, was also endowed with little success, and after a few months I was again on the move. A final attempt in this direction was made when I began an apprenticeship as a painter. My stepfather had given me the idea. But I also felt no motivation, here either. My passion, which I felt with each failed apprenticeship more clearly, lay somewhere else. And only there, in a place where my passion would have a chance to unfold, would I be able to find success. As much as I applied myself to the training—the most I could get out of my stepfather in feeling and praise for my painting work was a simple sentence: "I guess that'll do." I couldn't be satisfied with such mediocrity. There had to be something where I could rise up and use my abilities more productively.

The first thing I needed for this was freedom and independence. I noticed that I could not unfold my strengths under the burden of rules and directions. I needed the mental freedom to make new plans.

I started out small, delivering beverages in Gerlingen, a small town near Stuttgart. I didn't earn much, but I was free. There was no boss sitting behind me at the wheel telling me what to do. And that worked wonders in me. I loved my new found self-reliance, the feeling of taking control of my life. But I wouldn't get far in the long run in this job. The pay was only about 800 DM (German marks) per month. Too little. This was how I went from a fixed salary to a commission model. A colleague told me about an opportunity selling vacuum cleaners on a commission basis that would greatly improve my income. And so it happened. I discovered my business acumen at a company called Vorwerk. As the team leader, I was soon selling around 20 vacuum cleaners a week. But even here I was confronted by limitations. A working day only contained about ten effective working hours. That wasn't enough to achieve anything really big. And so I eventually came to multilevel marketing in the financial industry. This is where the potential was for me.

Everyone has a need for security. This is why people organize communities. They want to provide mutual protection from external dangers. It's no different with money. Security, in capitalism, means financial security, after all. Anyone who has had the rug pulled out from under them, has had all their savings disappear in one go, wants to be secured against an uncertain future. Anyone going for a quick buck without a secure financial foundation can be left without a penny in the blink of an eye. The financial crisis in 2007 was the spectacular proof of that.

Helping people to protect their money and at the same time provide sustainable profits became a calling for me during my educational training. For me, personally, it was a time that I am particularly grateful for. In those years, I learned to deal with the money and fortunes of other people in a responsible manner. Knowing that your money is in the right hands and being able to trust your advisers is the foundation required for successful investment. There must be openness and honesty between the customer and adviser. Unfortunately, you can't take this for granted in the financial markets. A situation that I find very regretful. This is why I have placed so much value in the most discerning customer care in my company from the very start. Nothing works without mutual trust, whether in the family or in business. From day one I have ensured that my relationships to customers are based on understanding their needs, which is very close to my heart. A lot of what we say, we don't say just with words. Our entire body speaks as well. Our posture shows how sincere we are. Our eyes are not just the metaphorical window to our soul, but the expression of our passion. My musical abilities may not be up to that of a "King of Rock 'n' Roll," but I'm convinced that one could see my enjoyment when I was singing. In that moment I was Elvis—100 per cent. Faith can move mountains. It gives us the strength to persevere and not get distracted, it gives us the self-confidence to make our dreams come true.

I loved having honest and open exchanges with others, long before my time in the financial industry. Whether among friends, in the family or later at work, I was interested in what moved them, what they were striving for, how they saw the world.

Once on my career path, I worked to set up the technical and financial foundation for independence. This time was the cornerstone on which I built my house and project, stone by stone, brick by brick. A house that provided security and protection from the wind and weather. But also a house in which other people would feel welcome. For the young Harald Seiz, that I was back then, this house was the promise of security that I had yearned for inside myself for so long. But it was also an exciting project, the challenge to find out what I was made of.

After some time in the financial sector, with all its ups and downs, I discovered a new side in me. I no longer wanted to be a little cog in the big wheel. I wanted to develop my own ideas and further the world with my innovations. Like the first European immigrants on the American continent, I was looking for freedom and independence. Just as they had once so laboriously crossed the Atlantic, I had to overcome my internal and external hurdles to find myself and my entrepreneurial energy. Their goal was the land of unlimited opportunity. My goal was to take advantage of these opportunities. Like the first gold diggers with their dreams of riches and wealth, I needed an exciting idea and the right touch for implementing it. I quickly realized that I had to step away from the beaten paths of my predecessors to be able to create something new. I could no longer stay in the rut of daily life, I had to find my own rhythm. I wanted to work more dynamically, more varied, and more creatively. This was the only way I could make the most of my talents. Creating the symphony of my life with all its nuances, instead of always playing the same old song—this was my mission.

In spite of all the comforts that go along with a well-organized daily routine, I became aware of one thing. What had been enough for me in the first years of my work life, security and familiar surroundings, had become a golden cage that I had to break out of. The desire to get to know myself and bring my own creation into the world was so great that I couldn't wait anymore to take the first step as an entrepreneur. I had changed. In terms of my character, but also through my business experience. I had gotten a sense of what customers wanted, of the gaps in the banking and wealth management markets. Filling these gaps was both incentive and challenge to me. I didn't know whether I would be able to develop a concept that would go beyond the existing assortment of investment possibilities. But giving up was not an option, for exactly that reason. I wanted to take on the challenge, because I knew that the usual investment products had weaknesses as well as disadvantages that I could take care of with my own concept. But I needed more than just a stopgap. I needed an idea that would set the market on fire. An idea that would ignite excitement in me and in other people. That would spread like brush fire because it was so well thought out and so impressively introduced.

Having a good idea is one thing, being able to demonstrate its value is another. How many revolutionary ideas fall into oblivion, how many concept papers disappear into a desk drawer because they don't reach the right ears. How would Marx have ended up without his supporter and publicist, Engels? What would an artist be without the gallery owner, museums or exhibits? A lone wolf that would vanish as fast as he had appeared. Long-term success is only possible with a presence in the public eye. And I knew that this presence required an attractive appearance and a company with a strong profile. But before I could think about the marketing concept, I had to work out the message, the performance, and the added value that my project would have for its customers. The big question was: How to safeguard people in the long term and in a long-lasting way?

My answer to this question was as simple as it was incisive: Gold. For centuries people have been fascinated by gold. Whether in sailor's tales of pirate's treasure or in legends of hidden caves filled with gold, everywhere in literature you find wealth and happiness expressed in terms of gold. Most people, when they were little children, heard the story of Rumpelstiltskin, who could turn straw into gold. A wonderful and wondrous idea. Gold has always been an object of desire, the most well-known symbol of wealth and prosperity. Yet most people have never even seen pure gold, let alone hold it in their hands.

Gold plays an immensely huge role in the financial markets as the most important and stable standard and as a secure investment. That's why it's hardly surprising that gold has a place in global economic policies. Since the Bretton Woods Agreement, gold has formed the basis of the US dollar value calculations. With the termination of the agreement at the beginning of the 1970's, this association was abandoned. Since then, our money has become one thing, above all: a fictitious promise made by the state. A promise that requires us to put faith in politicians and statesmen who we often hardly know and who are just as powerless against the momentum of the global financial markets in times of crisis as we are.

Gold was the answer. The alternative to blind faith in the obscure processes of politics and the financial markets that we are supposed to have. A means of securing customer stability in times of inflationary fluctuations and crises. An honest asset, whose only assertion is what is really inside.

My challenge was to take this realization and transfer it into a practicable business that would be able to function. An abstract idea is a good start. But I would only be able to burn with passion for a concrete project. A vision is important. But putting this vision into practice is at least as important. Setting up a business that would provide visible and tangible solutions to people, was the decisive stage that I had to reach. I was at the foot of the mountain,

23

unable to see the peak. I didn't know what would be waiting for me during the climb. But I was absolutely determined to make it to the top.

The experiences that I had had in my childhood were a great service to me at that point. Setting something up that would last, and was the embodiment of security and stability was a task that I had faced over and over again. In my first brainstorming sessions, I had laid a vague foundation, a draft with no blurred edges. Creating an architectural masterpiece was the real feat and at the core of my mission. I was aware that gold, in contrast to hard cash, did not depend on simple promises of the state of its value, but was continually being traded in the markets. How would it be to develop a product that would no longer be dependent on the good will of a government, but instead embodied its own value? The idea for the Goldcard was born with mini gold bars of 1 to 5 grams embedded—it has become the trademark of Karatbars. The company that I founded in 2011.

I am sometimes described as a financial sector rebel, because of this idea. However, I am not standing in front of barricades, armed to the teeth, smeared in blood and sweat at the forefront of fundamental political battles—as one might imagine a real rebel would do. I wouldn't want that either, because I don't want to force anything on anybody. People's interests are too different and their views about our world too diverse. But I do take some pleasure in the fact that there is a certain truth behind my rebel image. Because I also have a rallying cry and though I am not fighting against the system, I am trying to improve it with my ideas. I am realistic enough to see that I cannot turn the entire system upside down. To have success, you don't need to reinvent the wheel. Most great ideas don't arise in a vacuum. They build on what our predecessors have created, relating to the thoughts of the generations before us and ensuring societal advances. Every new invention is usually just a tiny step forward, but with many tiny steps you come closer to the goal. Often quicker than one thinks. One has to be patient, stay persistent and never lose sight of the vision. There is no point in delivering a rousing speech one day, if you're not prepared to carry out all the tedious tasks required on the count-less days and nights that follow. Being able to push yourself to the limit, even when it hurts, is an indispensable virtue. Like an athlete that prepares himself for years for a competition, an entre-preneur must give everything. The foundation and development of Karatbars was and is my Olympics, as it were. The thanks and encouragement from my customers are the trophies, the awards, and the gold medals that I have been continually striving for up to the present day.

In every corner of our life there is room for improvement. With Karatbars, I masterminded a small uprising against increasingly short-lived speculations. The overwhelming demand for Gold-cards showed me that I was not alone in believing in my idea. There are a lot of people, all over the world, who are a part of this idea, because they recognized the value and wanted to take advantage of it themselves. A single voice fades away quickly, but a rallying cry, sung out by a huge choir, rings on in the ears of all for a very long time. This choir could not be more diverse in its makeup. People are investing in Karatbars in over 120 coun-tries around the world. They speak umpteen different languages and have the most various of lifestyles and cultures. But they are all united in their dedication to Karatbars. This idea brings them together, ensures their long-term wealth and a financial founda-tion that will make it through times of crisis. The interest and sup-port that I have experienced with Karatbars is enormous. It shows me how convinced so many people are of this idea. Today, I meet with customers in all parts of the world and I am proud of the openness and enthusiasm that greets Karatbars. It's unimagina-ble, when you consider that we were still an unknown company as recently as 2011. But the passion shown for this project back then and today is the driver and guarantee of success for what Karatbars is evolving into. 100 per cent passion was what led me to found Karatbars, and the only way that I could win over so many differ-ent people around the world for this project. Because as different as our languages and cultures might be—the symbols of passion and commitment are universal.

I might seem a bit too old to pass as your typical rebel. But the decision to bring Karatbars to life, was grounded in almost three decades of ups and downs that I had experienced in my professional and personal life. Experiences that had given me the certainty that I wanted to make things easier for others, in the area where I was most suited. My determination to become independent grew firmer with each step, preparing me for the setup of Karatbars. Those who see the rebel in me should also consider the lengthy history leading up to this somewhat unique "rebellion."

Although I was not born to the life of a rebel, I can definitely assert that I viewed the world with an open-minded nature. I continue to experience great pleasure encountering people with an open and warm heart, as not only my little "Elvis" anecdote shows. That is perhaps my greatest talent and has led me to the message that I have taken from my own biography: "You must burn for it to be able to ignite the fire in others." Passion is contagious. And it's even more contagious when you radiate with passion. How else can you explain that tens of thousands pour into stadiums, concert halls and event locations to see their idols in person? People who do with passion what they were called to do are in the spotlight. They inspire. Whether in sports, music or in the world of business. The symbiosis of passion and the power of persuasion is self-evident. Those who discover what they are burning for inside, rarely find it difficult to convince others. Or at least to try. Even if it may not be a success from the start, with a bit of courage and the right presence, sooner or later opportunity will provide a chance to further develop oneself. We are not alone. Everywhere in the world there are fellow combatants with similar ideas of a better world. We all have our desires and worries, passions and fears. That makes us what we are, human. We have to learn to speak openly about it. There is no other way and I have come to see that in more than three decades of professional activity in my exchanges with customers. They also have their very individual needs. Developing a product that fulfills a part of these needs is my passion. And I am very happy that Karatbars is the right answer for many.

2.
"GREAT ACCOMPLISHMENTS COME TO THOSE WHO MASTER THEIR CRAFT."

Looking into the future is like looking at the sea. Your gaze wanders across the water until it comes to rest on the shimmering sunlight. A few clouds go by. The crystal clear view begins to blend in with one's own thoughts. With a vision of how the future ought to be.

One needs strength on the path to this future. The way is long and we need passion to go the whole way. It is the source of our conviction both within and without.

Vision and passion are the right mixture for success. They form the cornerstones from which your life's project can be pursued and accomplished.

However, there is a third quite fundamental ability which is necessary for success. That is expertise. You must be an expert in what you do. My former group leader in my first job as a self-employed salesman at Vorwerk is a good example of this. He was the best salesman in the entire company, a giant. Although the name Vorwerk was quite well-known at the time in Germany and it was relatively easy to get people to take part in a sales talk, my group leader was at the very top, selling 30 engines a week, and a great role model to me. But how does one become an expert? How can one become a sales genius like my group leader?

The bad news first: No one is born an expert. The good news: Everyone can become an expert. One only has to be aware of the multitude of different topics and subject areas that people around the world are dealing with. Whether the virtuoso with his instrument, the professional athlete with his game or the manager with his rational mind, they all are experts in their fields. To know everything and everyone well, to have a grasp of all subject matters, has become impossible in the complex world in which we live, today. While there were once polymaths, who became masters in several fields of study, today it is an illusory undertaking. While Leonardo Da Vinci was making groundbreaking discoveries in the natural sciences and simultaneously developing an artistic language of imagery nearly unsurpassable in its beauty and precision, today it is a huge achievement if one is successful in even one of those fields. Many fields have become so advanced, it can take years just to catch up to the present-day state of knowledge. Research has developed so that the tiniest of ramifications can be simulated. There are innumerable works and studies on every imaginable subject. Even though the Internet has made access to information easier, it doesn't necessarily mean that new abilities and knowledge will come running. The mass of information has become enormously dense. Trying to wield your way through the tangled undergrowth is a science in and of itself. Nowadays, almost anyone can give their two cents on just about every subject, which doesn't necessarily improve the quality of the information. The current debate on "fake news" has shown that not all that glitters is gold.

Life is finite and many simply lack the time to come to terms with all the topics that are currently making the world turn. There are family obligations, individual ambitions at work and many other things that take up our time in addition to pure curiosity. Anyone who overdoes it with their responsibilities and aspirations is rapidly in danger of running up against a brick wall. An excess of different responsibilities leads to stagnation. Nothing works anymore, both mentally, in one's own self, but also in the projects that one has undertaken. Instead of investing all your energy in one really important thing, you must suddenly divide your energies. One's drive is lost and the paper chase begins. When your organizational activities begin to take the upper hand, you know that something has to change. A single project sometimes only goes forward slowly and with tiny steps. Imagine what happens when many large projects are looming ahead. You then need either a larger team or a readjustment of your focus. I will show you now how you can do both.

Let's start with the right focus. It's important to ensure that your abilities and mental energies are focused on what the highest priority is. To be able to do this, of course, you first need to become aware of what the point of your life is and what your visionary mission is that you are prepared to pursue with your lifeblood and sweat. Once these decisions have been made, it's full steam ahead!

The earlier you take these steps, the better, and, therefore, the sooner you can orient your life in this direction. And by this I mean acquiring the abilities that you need for achieving your personal goals. Because one thing is clear: time is finite. If you want to produce a painted masterpiece, you must familiarize yourself with the fine arts. If you're looking for athletic success, you'll need to create a training plan and change your eating habits. Depending on what your goal is, the life you are leading must be conducted in the appropriate direction.

Instead of hanging around, without direction, it's time to give some shape to your time. With your goal before your eyes, you can start to create the structure, day by day, to become a master of your subject, in other words, become an expert.

A role model, teacher or motivator can be a great help, here. My group leader played a huge role in my professional advancement in the beginning. I learned the basics from the initiation to the closing of a sale from him. I was soon earning three, four and sometimes even 5000 DM per month. My professional development had an immediate effect on my salary level.

But there are only 24 hours in a day. The amount of vacuum cleaners that I could move in a day was limited. And even when my salary seemed generous in comparison to my work as a beverage transporter, I sensed that I was not at the end of my development and that I still wanted more.

An offer for a job in the financial sector came at just the right moment. Entirely in line with multilevel marketing, I was to set up a group through which we would distribute certain financial products. Even though I had already learned a bit about handling business transactions at Vorwerk, I felt like I was starting from scratch again in many areas of my new job. But considering the new possibilities that this job offered me, it was worth the effort. I was young, full of energy and ready for something new. And the impression my new boss made on me made my drive even stronger. He had a great influence on me. I thought: "That's what a successful businessman looks like." A dark green suit that fit to a T, and a white breast pocket handkerchief as accessory. Perfect! His outer appearance was the mirror of his inner posture. This man had charisma. I had sensed it since our first meeting. He was able to mesmerize people with what he had to say, but above all, with how he said it. He was my idol. Looking back, I am now aware of the importance of these role models along my life's journey. From Martin Luther King and Elvis in my childhood, to the group leader at Vorwerk in my early self-employment to this mesmerizing boss at the Society for Wealth and Investment Advice (Gesellschaft für Vermögens- und Anlageberatung, GEVA). There I also got to know Uwe, who would later become one of my closest friends. I admired him for his commitment and his enthusiasm for work, when he was taking care of his team. He combined a great nose for business with a sensitivity for social engagement. He has been with me through thick and thin and advanced to one of my most important companions in the course of my life. It was these people who inspired me and spurred me on to new achievements.

Saturdays were our most important work days. Every Saturday we made our appointments for the coming week. This was not always easy because we first had to convince people to stay in conversation with us. But we were ambitious and highly motivated. We didn't stop until we had at least twenty appointments. I made my first appointments among people in my circle of friends. With the help of an experienced employee, I analyzed which products would be the most suited for my customers. Life insurance was very popular back then. They brought in the highest amount of commission. But it wasn't about the money to me. The customers would also profit from it, so went the thinking back then. Because life insurance closed the gap between the current level of income and that at the start of the upcoming retirement benefits. Back then this was only about 70 per cent of the monthly salary that made life insurance the ideal instrument for covering the remaining 30 per cent.

With this benefits plan in tow, I made my way into the sales work. And quickly became one of the top executives. While many of my colleagues really had to struggle to get ahead, I was able to carry out my work with ease. I had some kind of style that appealed to people, that gave them the feeling that they could trust me. We earned good money, went to parties and out to eat. For the first time ever I felt that I could achieve really great success with the right work.

But the way there was long and difficult. When you carry through to the end, though, it's worth it. 100 per cent. Because following through is proof of one's willpower and conviction. This kind of stamina makes you stand out from the pack, who gave up before you. You are able to take your skills to a new level that will remain inaccessible to others. When an expert is sought, your name will come up. Special tasks require special skills. The smaller the circle of people who have the skills, the more in demand you will be.

Why do doctors and lawyers earn so much more than a waiter or hairdresser? Because they have invested a large part of their life in studying one of the most complex trades. The anatomy of the human body, the legal system of modern countries are so complicated that both are not accessible to most people. It requires stamina, great self-discipline and intelligence to learn these subjects. Not everyone is cut out for this. But in the end, you profit from your efforts. Not only because you earn more money, but because you also play a role in shaping the world with your talents. The proportion of lawyers among politicians is considerably higher compared to any other professional group. They are the ones that structure our life together, perched on the levers of power, bringing laws into our society. By becoming an expert in your subject area, you simultaneous multiply your opportunities to shape the world. Like a carpenter creating new furniture, the entrepreneur develops new business concepts. With your individual skills you can set things in motion to implement an achievement that was initially just a thought. However, you need clarity about what you want your achievement to be and what you need to be able to do to make it a reality.

I developed my vision in my work in the financial industry. The banking industry put out some feelers to me in my early years. I had grown up in turbulent family circumstances and yearned for security. Financial security was, to me, the basis of a stable life. With this idea in the back of my head, I stepped into a career as a finance and wealth advisor. I already knew then that my life's work would involve improving people's lives by giving them financial security. I wanted to develop an investment concept that would be able to withstand any crisis, so that other people would be able to profit from it. But at the same time I knew that I had to learn to master my craft. At the beginning I wasn't aware of how far the banking and monetary system had evolved. I had no idea about the needs of the customer, the work of a consultant or banker or the facts and figures in the financial sector. I first had to acquire all this knowledge. Faster than expected, I harnessed the necessary expertise. I began to deal with the diverse offerings and models of the banking institutions. As an employee working in sales at Vorwerk I had been thrown into the deep end, I soon came into intense contact with customers. From now on it was my task to convince people in direct conversation. I employed the body language that I had already tried out in countless stage appearances. I got to know an immense variety of people. Each one was special in their own way. Whether in the way they expressed themselves, in their expectations of what a good bank should offer, or in the question of what security really meant. Listening attentively, keeping my eyes and ears open regarding other people's worries—this all made me a more mature person, in my private life as well. Being able to react accordingly with the right recommendations and in the right tone required a great deal of adaptability and flexibility. Skills that I have profited from until this day.

I had to find out what was important for each person. In looking for the most optimal solutions I noticed something missing in what the banks and financial advisors usually had on offer. The portfolio that I was able to offer did not cover all of the customer's requirements and I had the feeling of having to lose out too often. Especially, in the segment of smaller, more secure investments with higher mobility, where I believed very important investment opportunities were missing. This problem was my chance. Because it was this gap that I would later fill with Karatbars.

The knowledge about the financial market and the practical experience with customers in sales were the greatest achievements in my vocational training. Here is where I got a feeling, day after day, of what was really important. Having the right presence with customers, a trustworthy manner and individually tailored products were the tools of my trade. I am thankful for this time, in which I could try things out and gain new experiences every day.

But even the most educational time has to end someday. In my case it was faster than I had expected, because after one and a half years the internal revenue came knocking on my door. The financial authorities wanted to know how much money I had earned. I had spent everything I had earned in the exuberance of my success and couldn't give them an answer. They made an estimate of my salary and because I hadn't paid, it wasn't long before I heard from a court bailiff as well. In the euphoria of my newly won independence, I had become naive and suppressed other responsibilities. I didn't have a tax advisor, so I now had to learn my lesson. Bit by bit I paid off the back taxes.

Like so many people in this world, I could hear the money calling and followed. Some colleagues had figured out that an insurance policy paid us a lot more in commission. Back then we were paid in so-called "per mille." While we only received 20 per cent

at the Society for Wealth and Investment Advice, the Stuttgarter Insurance company paid 35 per cent in commission. Almost
double in amount. I couldn't resist this option and went with the
others who had moved on. But instead of earning more money,
the opposite was the case. I entered far fewer closed sales in the
books than before: I was lacking in motivation. There were no big
sales days anymore, no group team training meetings or anything
like it. I was completely independent, could work when and how
I wanted. That was very comfortable on the one hand, but on the
other hand it made me become too comfortable, so that I began
to make a much lower turnover than before. Motivation should
never be underestimated.

When I realized that I had grown used to my tasks, I wanted to
look for new challenges. I had become one of the most successful
salesman in my division at the Society for Wealth and Investment
Advice. I realized what I wanted and that I had to try out new
things. I wanted to allow the potential that still lay dormant inside
me to unfold, at least in the long term, that would give me more
options, more opportunities for self-determination and would
require more creativity. Devising your own products and running
your own company with its own philosophy and strategy—those
were the arguments that turned the scales in favor of my decision
for entrepreneurship.

I stood up and banged my head on the metaphorical ceiling.
My abilities and ideas had developed further, my opportunities as
a sales employee were still limited. I had learned the tools of my
trade. Now I needed to put them into practice every day, to hone
them to perfection. I saw room for improvement in every nook and
cranny I looked into. The really great painters had not only copied
what their predecessors had brought to the canvas, they had also
developed their own style, their own language. It was exactly this
quality that made their work so unmistakable for all of eternity.
And exactly what also ensured that art continued to develop over
the centuries. My canvas, if you will, was the financial market in
its status quo back then. I knew how it functioned. I was aware
of what it was all about. But instead of going along with the same

old game of customer acquisition and the typical forms of investment, I wanted to put the stamp of my own ideas onto the world.

This would succeed after many years in further career stations with Karatbars. The Karatbars concept added a new facet to the financial world and expanded the choice of investment possibilities for customers. I'm proud of that. With this concept I didn't go by the standard playbook simply replicating what was already going on in the banking world, I gave it some fresh insights. Thinking outside of the box to break new ground—all this with the necessary background knowledge and always equipped with the tools of a competent financial entrepreneur. My color palette was my knowledge of the variety of financial products, my paintbrush was my skilled social interaction with customers. Thus equipped, there was nothing more standing in the way of a really great project.

At the same time, my step from sales to running my own company came with some requirements of its own. I had to go from being a cog in a large industrial wheel to being able to control the processes within this industry, myself.

I had to develop my own, functioning mechanism that would be up to achieving the goals I was striving for. In other words, I had to switch from being a knowledgeable financial adviser to that of an expert in my field. But what would my field, my supreme discipline, be? My fascination for gold which was already growing in me was the impetus to take a much closer look at the gold market. My discoveries were astonishing. They became the breeding ground for Karatbars. Gold's natural stability was what made it so much more attractive as a secure investment than almost any other product. This was even moretrue in the times of wild financial speculation and reckless fiscal policy making. The economic and financial crisis demonstrated in 2007 what you could lose when you put all your eggs in one basket. Gold is a perfect backup in emergency situations for any investor, whether venturesome or not—it's a safe haven when the seas become stormy and financial markets start to go crazy. Gold is independent of political decision-makers. It protects against cash devaluation such as in Argentina in 2002 when the debt crisis arose after their currency was no longer connected to the dollar. Independence and secu-rity are the keywords associated with the great advantages of gold. Advantages that have rarely been grasped up to now, particularly by small investors. They are simply lacking in suitable investment opportunities. Intensive studies, the precise observation of the market and an analytical viewpoint of world affairs, turned me into a gold specialist. I paid attention to all the aspects of this precious metal. From its extraction, composition, history and above all to its potential for investments. Bit by bit I developed into what I am today—a gold expert. In a nod to my idol Elvis Presley, a trade jour-nal called me the "King of Gold" on their front cover. Not because I could make decisions regarding all the gold reserves in the world. Even with a million-dollar company like Karatbars, I'm still far away from that. But because I had painstakingly taken apart every aspect of this precious metal for consideration, down to the tiniest detail. In this way I was able to acquire very specific expert know-how, which only a very small portion of humanity has at their dis-posal, if I may say so. I create new concepts with gold. Thanks to my knowledge, I know where the undiscovered potential lies. Like a pioneer on an unknown path, who charts a new land, bit by bit, I made my way through the thicket of the financial jungle. Always with the thought of making the advantages of gold available to my customers. Always striving to improve the world and enrich the financial market with previously undiscovered facets. I am abso-lutely convinced that the ability a CEO has to create something big depends completely on their expertise. It is indispensable for com-ing up with new ideas, building new bridges and spreading the wealth to everyone. If I am able to see one connecting idea in the experiences I have had along the way to becoming the manager of Karatbars, it's that one must master their craft in order to achieve great things. But just the craft alone is not enough. Understand-ing the basics of a profession are only the beginning of developing much deeper lying know-how and skills.

Our world is highly specialized. The problems that we worry about are becoming much more complex. With the scientific find-ings gained over the past few centuries, a sense of the complexity of life on this planet has also arisen. We are able to examine things to the tiniest degree. Phenomena become visible under the microscope that no one could have imagined even existed before. With binoculars we can look into seeming endlessness, and see that we are only a small part of this universe. We have calculated the pathways of the planets and their distance to the earth, millions of miles away. What we know is only a tiny part of the big picture.

Today's findings are based on the knowledge that people worked out centuries ago. Just tapping into a portion of this treasure is already a huge challenge for each individual. Shining a little light in the darkness makes the world at least a tiny bit better.

To be able to achieve that, there is nothing better than specialization. Those who discover their passion need to follow its path. Our passion points us in the right direction and gives us strength along the path to expertise. For me it was the fascination for gold. For an atomic physicist it is the enthusiasm for the tiniest particles of our universe, for the painter it is the composition and structure of a painting. It doesn't matter what it is—passion is, and will remain, the engine empowering our performance. Expertise is the tool that enables us to complete this performance. The list of the various human callings is unbelievably long. It's the expression of a world with immense diversity. This world wants to be discovered, explored and created.

The expert represents a small part of this diversity. He delves into the depths instead of remaining on the surface. He challenges what seems to be obvious instead of simply accepting what appears to be there. Always searching for something new, he sees opportunity in change. With his abilities he helps other people to make the right decisions. When everyone answers their calling, they can all profit from one another. The individual can bring in their expertise for the common good. Whether as a researcher, creative artist or financial advisor—everyone contributes their share to society. Talent and intelligence are the raw diamonds and they need to be polished. They radiate with luster and perfection, when talent is used in the right place and is continually encouraged. It is up to every individual to polish their skills and implement them where they have the greatest use for society. Hardly any other political project embodies this idea as well as the European Union. By getting rid of internal borders, people, regardless of their nationality, can move within Europe to the place where their services are most in demand. In times of globalization, it is no longer appropriate to think within a small framework. Good ideas have the potential to change the entire world. Thinking big means thinking European, and it also means thinking and acting globally.

A passionate entrepreneur does not only help society, he is the embodiment of his burning passion. The incarnation of a great idea is the globally networked manager. He does not only have himself, his company or his country in view, he is also looking at other people, continents and the world. This is how personal fulfilment and the common good go hand in hand.

It's about happiness gained through expertise, because one is able to do what one is best at and at the same time receive a reward for it—in money, but also something that is much more important, gratitude, recognition and respect.

In order to achieve this symbiotic state, expertise is not enough. What good is knowledge and experience if one is not able to share it, if no one knows about it? Even the greatest scholar in the world doesn't benefit the common good if he doesn't proffer his knowledge, make it known or make use of it.

A bookworm, who takes in everything but doesn't achieve anything with it, remains without an affect on humanity. What did the famous thinker, Karl Marx once say? "The philosophers have only interpreted the world in various ways, what's important is to change it." No quote could be more to the point than this one. It is about putting into practice what one has learned. Because even though some theories don't work in practice, progress is only possible when you take some risks. "Trying is better than studying"—ideas of genius begin to unfold and show their usefulness according to this motto.

My colleague at the Stuttgarter Insurance company, Tino, and I soon opened up our own office in order to be able to give even better advice to our customers. One day a man came into our office and told us about a new investment opportunity with a 10—15 per cent rate of return per month. We rubbed our eyes and were skeptical at first. So we decided to go and look at his offer in person and drove to his company headquarters in Switzerland. The advisors there talked about stock market business that had achieved up to 500 times the profit in only two hours. Tino and I looked at each other and nodded: "Sounds good." Although we didn't yet know all the details about it, the concept as a whole sounded plausible. First we invested, ourselves. Tino threw 50,000 DM into the pot. I threw in another 30,000 DM. As we allowed the investment to run, it quickly ran true to its promise. We received 12 per cent interest per month. 5000 DM without doing a thing. That was unbeatable and quickly made the rounds. I told our customers: "I don't have to work as hard anymore because a portion of my income comes from my investments." That made them curious. Even old friends got in touch and asked about this investment. The Swiss company promised us 3—5 per cent commission for each new customer. A super deal! And it went like clockwork. We brought our friends to Switzerland and they paid their cash into an account of the company there and we got the commission. The whole thing spread like wildfire. Every new customer brought further interested parties to us. We were young and believed in this model. But a small amount of doubt remained. Wasn't this too good to be true? Within two and a half months I had brokered over 1,000,000 DM. I had never earned so much money in my life. I bought a Porsche, rented a huge apartment, threw big parties. It was like life on another planet. Our final doubts were gone when the company boss of our partner company bought the Brabham formula 1 stall. We thought: "Now he is in the public eye under his own name. Nothing can go wrong, now."

Unfortunately, we had to learn a hard lesson. First, they said there was just a slight payment delay. Not to worry, they said, there was a small problem with the company software. We waited, believed and hoped that the payments would come. Until the company suddenly disappeared into nowhere. No answer, no reply, no contact anymore—nothing. Our customers were angry. And they weren't even our customers. We had simply acted as broker. They had practically begged us to do this. When things were going well, we didn't hear a peep from them. Not the tiniest sign of gratitude. But when everything went downhill, they came running. This went so far that they were calling us daily with threats to our lives. The people wanted their money back, otherwise something bad was going to happen. I felt extreme pressure, was intimidated and finally saw no other way out than to lay low in a small hotel for a while, just to be able to start thinking clearly again. What could I do, to make these threats stop? The company and the money were gone. There was no one to shake down. I finally decided to go on the offensive with the problem. I thought a good offense is the best defense and went to each, individual customer to speak openly with them. One has to deal with problems proactively, talk about them openly, be honest, even when the truth is uncomfortable. And the result was startling. Apart from a single person, everyone gave me a chance to talk with them. I was so relieved. Even more: My entire emotional state had been transformed. Because I had successfully mastered the most difficult situation up until that time in my professional career. But I was not yet over the hump. The problems carried on. Because now the district attorney was on the case. The people wanted their money back and had brought on a lawsuit against me. I did not feel culpable in this case. At the end of the day, these people were not my customers. They had almost forced me to get them in touch with the Swiss company. No one had any evidence against me. Nevertheless, on the day of the court case I was almost beside myself. What would happen? Would I be able to convince the district attorney and, above all, the judge of my innocence? At this moment my career was hanging by a thread. I found myself sitting in the courtroom again: a room decked out in heavy, old oak wood, perpetrators of all complexions had sat witness, here: thieves, fraudsters, murderers. I didn't belong here and yet here I was sitting in just this place. An oppressive feeling came over me—the long hand of the state reaching out to get me. After the criminal charges were read aloud, the district attorney turned to address me: "Mr. Seiz, as an investment adviser, you must know that no one can receive 15 per cent a month." But I had the appropriate answer ready, and responded: "Mr. District Attorney, you should have seen who was sitting around that table, putting their cash in bundles of 10, 20 and sometimes 100,000 DM on the table. It was lawyers, tax advisers, police chiefs, judges. All of the people believed it back then." That was persuasive. After this, the district attorney said only: "No further questions."

In the end, I got out of it with just a slap on the wrist. But more important, I had learned something of significance. Of course, you can never know if a business transaction is going to work or not. It's much more important to pay attention to your gut feeling. The whole time I had been working with this Swiss investment opportunity, I had had a funny feeling in my gut, a feeling of uncertainty and unease. And this feeling turned out to be confirmed, in the end. Being aware of my own gut feelings would influence all future decisions. Although there was not much left from what I had earned in commission, I was definitely richer in experience. Even if you are never 100 per cent sure whether an idea will function, you should always take some risk. No entrepreneur was ever successful by avoiding risks. The relationship between return on investment and risk is what's important. No risk, no fun. And if you have a good idea, and have mastered the tools for implementing it, you should try. First of all, because you should expect that your own expertise is more than what many people master in skills. Second of all, because you will only be able to go further and gain your experience by going out into the world and putting it into practice. Consider the genius, Isaac Newton, who was the first to come up with the theory of gravity when an apple fell on his head. He felt the pain of gravity from his own experience. This feeling was the beginning of research on gravitational forces. It can be such seemingly banal things that happen that give us brilliant ideas. Only those who observe the world keenly, and transform ideas into reality, get closer to success. If a good business idea has begun to take shape and the first few steps are made in the implementation, some important questions will follow. How will others hear about your idea? How will the customer know that you are there? How will they know that someone is out there that could have a solution to their problem? A solution that is built on a solid, theoretical foundation. How will he know that you have provided a solution that has not just come from a feeling, but has been worked out rationally? This is where the most essential thing comes into play for every entrepreneur, and that is the network. At the very latest after the business concept is finished, the market has been analyzed and the first customer base has been built up, everything will begin to revolve around the question of how the business can grow. New circles of customers have to come and new business partners need to join. Ideally, a win-win situation results for all those who participate. The customer profits from their new product, the business partner from a new cooperation, and you get new opportunities to make your business idea public.

A customer who is convinced of this idea will tell friends and colleagues about your product and a good business partner will bring new contacts along for you. The more well-known your company is, the more the demand for your product will grow. This simple mechanism should always be at the top of your mind.

The situations where you will need to be patient will not be rare. Because one thing should be crystal clear: building up a network is hard work. You are not alone in this task. Every day the telephone lines are ringing off the hook, email servers are filling to the brim with contact requests from well-known trendsetters. Getting noticed in this crowd is an art. Your potential new customer doesn't know what you could mean to him and will try to wriggle out of any conversation with you. You need to stay persistent without being intrusive. Sensitivity and tactfulness are important to find the right entry into a business relationship. Stay calm and be patient, even when the tone at the other end of the receiver might seem a bit rough. Often, a rejection is a result of the ignorance of your counterpart on the phone. He doesn't yet understand the advantages that a business partnership with you could bring. In today's world, where people are confronted with advertising on almost every street corner, a skeptical reflex to an unknown contact request is all too understandable. I would probably not react any differently.

But this reaction is not aimed at you or your proposition. The recipient cannot possibly know what he is missing. He is guessing and applying the general rule to his individual situation. And it's exactly here that you should begin.

Because every new contact is a chance for a successful business relationship, which both sides can benefit from. Sometimes over a period of years, if not decades. The setting up of a network is not advertising for the great masses, it's a selective process which makes the most of business potential. You are not just any caller, you are a businessman with concrete proposals and a unique vision. You can't expect your counterpart to react with this perspective right from the start. But that should not prevent you from seeking this contact.

The foot in the door is a personal approach. By looking for similarities it's possible to rise out of the mass of disjointed requests. A personal approach makes your genuine interest in working together more clear. But much more important: the person you are contacting feels that you are talking to him, personally. To achieve this, it's worth doing some research before you make contact. What can you find out about the other person, what have they achieved so far, what's their lifestyle and what are their interests? The answers to these and similar questions will help you, immensely, to quickly move from just getting to know someone to a broader basis of trust. And that is the goal. Because whether we're talking about your private life or business—nothing works without trust.

Along the way, this approach gives you the chance to highlight your vision and the passion you show for the project that you are presenting. In all honesty, nothing is worse than a boring conversation opener à la "I just wanted to or I thought I could." Show your enthusiasm right from the beginning and try to get a reaction from the person you are talking to. It's better to be short and sweet and to the point than going into the tiniest of details. Details will come later. Now is the time for persuasion.

Here you can work on a particular style, a type of standard repertoire for your first pitch. Introduced in an easy to understand way and containing a little anecdote—this makes your enquiry much more appealing. Draw out a picture of your plans and put the accents on your core message. Adapt your intro depending on what you think could be particularly interesting for your contact, from your point of view. In the first few seconds it doesn't really matter exactly what you say, the "how" is what it's all about. How you say something, the sophistication that you are able to convey. That's what's important in reaching the person you're talking to.

Here I like to think about Martin Luther King. Why did people listen to him? He had the talent of filling the things he talked about with life and feeling. He was a master of postulating his passion and in this way was able to persuade his peers at the time. He began as a simple priest in a little community and became an icon to an entire movement by the time of his early death. Martin Luther King was a master of rhetoric. Style and speech were inseparably connected for him. It is exactly this attitude that every businessman should learn from him. One can only profit from this. When you set up your network, make use of all the stylistic tools you have at your disposal. You will soon notice what works and what doesn't. You can refine your instruments and develop the qualities of a true "networker."

But what exactly are the levers and switches that one needs to put in motion when trying to get into contact with someone new? First of all, the voice. It gives away a lot about people. Because when one speaks, one is not only conveying the meaning of the individual words. There is another message that is emanating from the voice. Another level, that is much subtler than the pure content. Nowhere else is that as clear as songs in music. Whether you understand every word of the lyrics is not that important. It's about the voice of the performer. Does it sound strong and solid or soft and weak? Does it sound stressed or relaxed? Whether it's a song or a conversation—the criteria is the same.

Before you speak it is, therefore, advisable to relax your voice. Concentrate on the central theme of the upcoming conversation and think back to positive memories to give your self-confidence a little push. Believe me, that will be audible, automatically. Your personal state of mind comes through in the sound of your voice. And you depend on your voice to convey your conviction. Your voice transports your mental state to the outside world. It is all about how you can best take advantage of this outward expression. Mastering this skill will transport you to the fast lane of success. It should be part of the repertoire of a top manager. Because only those who can speak appropriately to their fellow man, move them and motivate them, will be able to establish a profitable network of partners and customers over the long term. That is the goal and the right implementation of the voice is the means of achieving this goal. Take your presentation to a whole new level by using your voice effectively.

However, the palette of persuasive possibilities has not yet been fully exhausted. It only becomes really exciting when the first contact is happy and a personal visit to them has been arranged. Because in addition to the acoustic aspect of the voice you now are confronted with the visual.

Body language—this term hits the nail on the head. Because you do not just speak with your voice, you communicate with your entire body. If I'm standing up straight, I demonstrate sincerity.

If I make some gestures, I strengthen the content of my message. There is a reason why generations of actors have practiced this art of using body language. The voice is not used at all in pantomime and yet we still understand exactly what is meant.

An example from real life situations is the handshake. It is often the first sensory impression that we get from the person that we meet. I usually notice as soon as we shake hands what I can expect. Is it assertive and strong? Does it show some presence?

Is it somewhat bashful, maybe even fearful? This little example shows how important the smallest gestures can be in forming the picture that we generate of the other person. An open and honest stature shows an open and honest frame of mind, a focused look shows an awake mind. It's not without reason that the famous saying goes: "The eyes are the mirror of our soul."

What eventually becomes routine for every single CEO, was probably once still unfamiliar, initially. That doesn't mean, however, that you should not try it at all. Quite the opposite: "Practice makes perfect." The more you consciously make use of your body language to persuade new contacts in personal conversations, the easier it will be for you in the future. No one was born a master of rhetoric. After many years of trying out and practicing in real life situations you will gain this new skill. Use every opportunity that comes along to practice.

In the use of body language remember, especially: "Trying is better than studying." Body language needs to be felt. It comes from within us and is the expression of what we want to convey. Your individuality plays a huge role, here. Skillfully putting this into practice will become your greatest trump card.

Elvis Presley did not copy his performance style from anyone. He developed it from show to show until he had made it inimitable. Everyone is different. That's why every type of body language is also individual in its own way. You just need to have the courage to show it and use it as a means of conscious persuasion.

An ideal place to try this out can be in the surroundings of your friends and family. Your friends and the members of your family know you and will quickly notice when you seem unnatural or artificial. You don't need to be an actor to reach other people. It's sufficient to simply learn to express your style and character authentically and self-confidently to the outside world.

For me it was the performances at talent shows that gave me a feeling of how I could give my personal style a special touch. Naturally, I was nervous to begin with. Uncertain of how I would come across to the audience. But I soon noticed that the important thing was to stay true to myself. With this inner certainty, I lived to the fullest on stage and learned to be persuasive through gestures, posture and movement. Because even though I was not a bad singer, my voice was not of the same caliber as an Elvis Presley. But it wasn't about Elvis Presley anymore. It was about my performance and I had been able to improve it with each performance and rehearsal. That was my path to persuasive body language. And whether on stage, in personal surroundings or in business—persuasive body language can be used everywhere. It's not sorcery.

Once one knows how to be persuasive both personally and professionally, the question naturally remains of who one wants to persuade. Who is this person that I've been referring to so generally as "customer" or "counterpart"? Let's shine a little light into the darkness and open up this mysterious black box.

Because before I can be convincing in my body language and verbal expression, I need to know who it is that I want to convince. Not everyone is going to come into question for the product that you have developed or the service that you have to offer. This is why the circle of potential customers and business partners should be selected with precision. Filtering out the right contacts for your business early enough, makes your work later a lot easier. Otherwise, you run the danger of investing time, effort and energy in something that isn't worth it in the end. Wasted love? Never again!

The first step in networking is the selection of the right contacts. Who might be interested in what you have to offer? Who can help you get ahead? With additional contacts, good suggestions or a beneficial partnership. You should also ask yourself what group of people would be especially interested in what you have to offer.

In the case of Karatbars, the pool of potentially valuable contact people was very large. Because both private savers as well as large investors would find a safe haven in gold investments. Anyone trading in the financial markets or working through a brokerage firm needs a certain amount of financial security. For an international business, it needed to be a broad network of partners who could represent the idea of Karatbars around the world. These people had to be competent when it came to gold and be eloquent in their personal style. That much was clear. Here we had set the foundation for the right contact base. We had developed the criteria to find those people who would be essential for getting Karatbars ahead, from a sheer innumerable mass of human beings, from the about eight billion inhabitants of this planet.

In the same way a manager must be competent in making important decisions, one must also be able to be selective when it comes to who should belong to their network. A well-maintained network that has been built up in a targeted way can help carry a company through difficult times. It opens up opportunities for stable growth. The true value of an idea really begins to unfold when it is further advanced by the right people to the right people. This is why "networking" continues to be propagated in the business world.

From my own experience with Karatbars, I can definitely say that networking is really worth it even though it requires a lot of patience. In some industries, a good network is a kind of survival insurance for the continuation of a company and, therefore, something absolutely indispensable.

Just think about the advertising industry. More than a few deals are made because of glowing recommendations from the right person, or taking advantage of the old boy network or setting up a new cooperation. Company events, meetings, conferences—behind all these event formats lies the ambition of making new contacts from which you later profit. Especially, when you are just beginning to network, these types of events offer high quality possibilities to get a foot in the door of the business world.

Casually, and usually in pleasant circumstances, you can draw some attention to yourself. Use this stage. Give them your best performance. Present your ideas and project plans. Those who give a lot, get a lot back. Only those who are brave enough to take the first step can expect others to do the same. Before my first show, both as a singer and as the CEO of Karatbars, I was pretty much unknown. I was an unwritten piece of paper and was under everyone's radar. No one knew the passion and persuasiveness that I had inside me. That all changed in a heartbeat when I jumped over my shadow and out onto the stage. A few minutes in the spotlight are often more valuable than numberless wasted calls. With the right touch you can set off a positive chain reaction that will lead to your getting into conversation with others. Ideally, you won't have to go looking for the people that you need anymore, because they will come to you on their own. First of all, because they now know that you are there. That is as simple as it is true. And second of all, because you were able to convince a large group of people.

Naturally, you need a big helping of courage to stand up in the front row. And you'll need some elbow grease more often than not. But the courage that you show here will pay you back a thousandfold. Even practiced show masters and entertainers get stage fright when they have to go out in front of an audience. Even Robbie Williams, who has innumerable tours behind him and thousands of fans, is full of butterflies before going on stage for each new concert. It's about how you deal with the nervousness. It can block you—who hasn't had a frog in their throat at least once? Or it can push you to your best performance. An adrenalin kick. Extreme sports thrive from this phenomenon. The unbelievable energy that is set free in the moment, is the wind under the wings of a success-oriented entrepreneur. These moments give you self-confidence and get you publicity. A dynamic combination that you will hardly find anywhere else.

Once everyone is talking about you, you can make a lot of new contacts. But beware! Quality is better than quantity in networking as well. A few quick and half-heartedly exchanged business cards will not get you any further ahead and will land just as quickly in the garbage can. That's why it's important to set priorities. Who is really important? Who can push your project forward? Do you need a network of partners that is as wide as possible, or a few specialists for a very particularly area? With the right answers to these questions you will quickly know what contacts are worth the expense and who you should leave in the dust. You have no time to lose, only time to gain if you learn early enough to work with the right people. That is true, by the way, for both sides: you and your business partners.

Sure, having lots of customers achieves publicity. And new orders, quickly. But it is at least just as important to make sure you've got the right people in the boat to be able to handle the demand, optimally. Because what else is going to give you viral resonance, if the high quality service or offer doesn't fit? Mostly, just a lot of frustration, for you and your customer. A good network ensures a healthy balance and grows with the company.

Each contract brings specific advantages. You need to put your network together carefully to make up a unified whole, like a puzzle. The individual pieces of your project need to fit together so that it functions. A large part of this puzzle is the network. Another one is your team.

Getting a big pack together is going to be a lot of hard work. But as a lone wolf you won't get very far. Everyone has their limits. That's why you need the right team. Everyone has their task and everyone is important in their own way for the overall concept to become a well-functioning organism.

One only needs to look at one's own body. We are made up of uncountable microscopic little cells. Each one with different information, different tasks, but all equally important. The same goes for the inner organs. No one can survive if the heart stops beating. But the liver, brain and kidneys are just as important for survival. The wonder of the human organism, is only possible when all parts carry out their role together. The laws of our body apply in the same way to corporate organizations. A company is also a living organism, in which every part has a role to play. Only then will success come, only then can it make progress. But the blueprints are different for every company. While the human body is always made up of the same organs, every business requires different resources. Divide the tasks so that everyone does what they can do best. And above all, that only things are done that are relevant to the business. That includes recognizing the potential that is in every employee, that can benefit the entire project. Is there anyone who hasn't heard of the young trainee who is asked to make the coffee? I say: enough of that! A trainee doesn't come to learn to make coffee, he comes because he's committed to helping the company. He wants to show what he can do, make a contribution and learn new skills. It's the CEO's job to find the right place for this commitment to be developed.

Many people believe that success is simply the result of a "sharpen the elbows!" mentality. Of course, assertiveness is sometimes required. However, it's essential to have an eye on the individual people that every entrepreneur requires in order to be successful. Without this ability, no well-functioning team can be built. And without a team, great success will remain a utopian dream. Your own team, if you will, is the most important network that you have. Those are the people who are currently working to achieve your vision.

You only need to take a look at the history of humanity. How could we get this far? We were once a mass of natives hunting out on the steps, today we communicate over thousands of miles and in the fraction of a second. In the meantime, we are able to create meat artificially, predict weather phenomena and delineate outer space. The achievement of all these milestones is due to an organizational masterwork: modern working practices. In which human beings no longer have to fight on their own for their basic survival—they can now turn their attention to discovering new things. Because there were farmers who ensured that other parts of the population had enough to eat, science and entrepreneurship could be established to drive human advancement forward. Up until today, companies have taken on an important role in the financing of research.

What took place in the course of human development is also indispensable in the setup of a modern company. Only when one part of the employees are taking care of daily business, can another part take on the work of long-term planning. Only when regular bookkeeping and customer service is guaranteed, can other departments work on entering new markets. And good networking takes time and energy, that you only have when you've got a team working in the background that has your back. The first network of a company are the employees themselves.

A company is a living organism. An organism that is constantly changing, in accordance with its nature. New competitors stomp on your plans, employees go or new ones join, technical developments continue to advance. All of this affects your own project. One needs to be aware that networking is not a one-off thing, it's not a bullet point that you can check off your to-do list. Networking is an omnipresent process. Because just as your company is changing through its growth, developing new strategies, refining its profile, the business world around you is also changing. Managing your contacts is a continuous activity which requires two things to happen, above all others.

The first is keeping a continually critical eye on what is going on. Not a few people assume that when something has begun to run well it will continue that way forever. That is a fallacy. The world is constantly in a state of flux. That is more than ever true in these times of globalization and networks. Who can guarantee that your products and services will still be needed in a few months? Who can be sure that there are no competitors lying in wait around the corner? Right. No one. And definitely not you, either. The market is in a constant state of development. The best security for the future that you can bring about is having a critical and realistic view of your own company and the developments in your environment—and that also means your network. As important as dreams and vision are at the beginning of your business development, you want to establish your business for the long term, and should not shut out the changes going on around you. A good network will help you with this. It reacts sensitively to changes. Like a seismograph, that registers an earthquake that is slowly getting stronger, your own network will provide you with the signals of change that is on the way. It's up to you to recognize these signals.

The second thing is awareness, not just in terms of your external network, but with regard to the inner circle of your own company. The advantages of a functioning team are obvious enough. But it is not a God-given thing that you will constantly profit from these advantages. The permanent state of carefreeness that is achieved when many people work well together cannot be expected to exist at all times. Because people change, their plans and motivations vary. Employees can reorient themselves, move on. They have to be replaced. Some may seem to be missing the right engagement. Never shy away from being honest. Only in this way can losses of efficiency be avoided. These are chances to clarify your visions and goals.

If you are not able to be honest with your employees, you are not being honest to yourself. Only by being open to criticism can you quickly and creatively adapt to the constantly changing world around you. If you take this knowledge to heart, you will have all the cards in your hand for setting up a company that will be secure in the future. Both in your own network as well as in the target market.

Every change brings some stress with it. It is not always easy to accept that circumstances have changed. Especially, when things are going worse than expected, it's easy to get upset. But this is exactly the moment when you need to make well-thought- out, balanced decisions to get yourself out of a crisis. How do you handle business partners who go over to the competition? What do you do when you no longer have faith in an employee that once served your company so well for many years? How do you convey to everyone that resources need to be newly redistributed? All of these are questions for which there is no easy answer. But there are guidelines that will help you make good decisions. It doesn't help to do nothing when conflicts arise, and just hope that it will one day take care of itself. Usually, it just gets worse. The consequences will become even more serious. Decisions that you have put off will catch up to you sooner than you think. In the worst case, they will overwhelm you when you least expect them.

It's about finding the right point of time to free yourself from firmly held illusions and face reality. Even though it can sometimes be very painful. In your private life as well, no one seriously expects you to hold on to a relationship where love and affection has gone. There are times when having an exit strategy is the only correct solution.

Taking radical steps requires a lot of energy. But once carried out, they open up quite new possibilities for creating your project anew. Accepting that nothing will ever remain the same enables your thinking processes to have more flexibility which are indispensable for a successful CEO. You have to keep your finger on the pulse of the times. Otherwise, time will pass you by like an express train. Don't miss this train, jump on it as long as it is still possible. The best example of this is digitalization. There is no point in ignoring change. It will catch up to you sooner or later. How many companies have waited too long to take part in the digital revolution and how many others have profited from it? In order to belong to the group of winners, a CEO must see the potential in new technologies instead of fighting against it. Cost and effort should be oriented towards the future and not towards the past. You can design the future, the past remains the same. Potential lies in the future. In order to call up the former, it might be necessary to completely reorganize your company. Think of an automobile factory in which machines are now taking over complex assembly processes to supplement the work of humans. You've got to deal with the fact that there will be difficulties in this new order. But holding onto the status quo will be fatal. If the company hits the wall, no one benefits in the end. As the CEO you have the responsibility.

But how do you manage to set up a team that can work successfully over the long term? It's not enough to simply roll the dice and hire a few people hoping that they will somehow make things happen. You have concrete goals and in order to achieve them you need great work performance. Just having a lot of people who work together doesn't automatically mean that they will also be productive.

Work in groups comes with some traps. Some will be too quick in trying to be successful, others will always push their work onto others. As a leader, you need a lot of sensitivity in getting to grips with the group dynamics before push comes to shove. This requires some experience, but there are a few simple tricks that will increase your team's performance. What's most important?

Make sure that you only involve people in your project that are convinced of the relevance of their work. That sounds banal, but it happens more often than not. How often do you hear about the employees who are only interested in getting their salary paid in at the end of the month? Of course, nothing works without fair pay and financial incentives. But that should definitely not be the only motivation. Those working in your team have to bring more than just the desire to get profit for themselves. Here I am returning to the topic of passion again. Only those who radiate passion for that which they do can also be successful in the long run. That's true for every member of the team, regardless of the position. If there is no passion, the motivation will also soon be gone. And no one wants to have an employee lacking drive in the office. Definitely not when you're the one paying.

That's why your goals also have to be the goals of your team. Making this clear as early as possible will set the right standard, which you will need often enough in the jungle of personnel recruitment. Because fair pay is one thing, but believing that your work is meaningful and working towards a purposeful goal increases the willingness for high performance and much, much more. Those who only do something for money become dissatisfied more quickly, and seldom become rich. Mostly because they never do things with the same enthusiasm as someone who cares about their work. Money cannot replace passion. And without passion, even the best of ideas is doomed to failure.

A team only comes into existence when everyone feels a part of it. Processes and skills have to work integrally together. You don't need a collection of lone wolves, but a group that sticks together. Team spirit only arises when each person complements the other and is willing to take responsibility for themselves and for their colleagues.

A team can perform much better when each individual recognizes that he is an important part of the successful whole. Not only by fulfilling their own tasks responsibly, but also by being flexible to help the others, give advice to colleagues and take on extra tasks for a short time when difficulties arise.

Problems at work should be spoken of openly in the group as well. Because five heads are usually better than one. In this way, solutions can be found by the collective for individual problems. This increases the feeling of belonging together and strengthens cooperation in the future.

We all have our problems and we all have our defects. What's important is how you deal with the problems. If you try solving your problems on your own, you risk going off on the wrong track and irritating those around you. If you inform your team, on the other hand, let your group know about your difficulties, everyone will be up-to-date and can deliver the help needed. Only through sufficient transparency, will you have the opportunity to create work processes that can be adapted flexibly to the needs and particularities of your employees. This is what the team is there for— to help each other and to learn from each other, by taking what's positive from each other.

This is why a successful manager needs interpersonal skills in addition to professional skills. Complex issues require clever brains, but sustainable success only arises when these clever brains can also work well together. Everyone has their own character and their own family background and their own special qualities. There is no point in trying to make a lot of changes. Certain things might be able to be changed. But that should not be the primary task of the business environment. Instead, you will get a lot further when you simply recognize and accept the truth. Recognize that people are different. There are advantages to be seen and taken advantage of, in this diversity, because they open up a different viewpoint. This is a part of the art of successful personnel management. If you are able to be respectful of other people's differences, you can expect acceptance yourself. When you recognize other people's problems, you should support them in a targeted way. Overcoming weaknesses together using strengths together—that is the way to march towards fruitful team work.

Communication is the top priority along this route. No one can read another person's mind. Speculating about the feelings and thoughts of other people usually results in more disquiet than clarity. Rumors start and what's really important, the work that needs to be done together, is pushed into the background. To make sure that this doesn't happen, it's necessary to maintain an environment that is conducive to open conversation.

An open dialogue is sometimes more healing than the most well-written regulations, regardless of how persuasive they are. Sometimes just talking about a problem gets rid of many blockages. An open and honest talk, whether face to face or in a group, is the door opener to a new level of quality in your team work. A CEO should instead push for an open exchange between employees. Everyone knows where they stand and what they need to do. That enables the certainty that you are working in the right place, and ensures self-confidence and improves people's ability to communicate and handle criticism as a side effect. Qualities that will pay out, in the truest sense of the word, later when in conversations with customers.

Have you already laid the foundation for your team and now need more human resources? You should go on the offensive to discuss this. A comprehensive analysis, of what competencies are required, what type of personality is necessary only works when the opinions of the rest of the employees are considered. The foundation of successful work lies in the makeup of the team. The small things and daily work processes can be corrected but complex skills aren't internalized overnight, the human character is not made of rubber. Personality and professional competence are the decisive factors in the required profile of new colleagues. You get the best results when you know exactly what you want. And when this "knowledge" is based on an excellent analysis. Don't forget that our thinking factory, our brain, is the greatest treasure that we possess. Use it. Use your common sense. And the common sense of your team. Big decisions should not be made in a back room, but spoken about openly. One of the most important decisions is making time. Because the best ideas don't just come floating on to your desk, they are developed in every exchange, discussion and consultation. You won't be able to take every aspect and every possible scenario into consideration when making your decisions, of course. However, you shouldn't rush into giving premature instructions. Sleeping it over a night and paying attention to your gut feeling can sometimes work wonders. How you make your decisions influences the quality of the decision to a great extent. At Karatbars we have developed our own concrete processes for making decisions, that both consider the input of the entire team as well as enable corrections and feedback. The right course of action for a firm must always be made afresh. The CEO, no matter how experienced and able he may be, is like the captain of a ship that depends on his entire crew. Because in addition to his compass, it gives him the information that he needs to do the right thing. That's what democracy is all about. Only when those, over whose heads a decision is being made, have an influence on the politicians making the decision, will they accept it. Do you want to have your decisions respected in your company? Then integrate your colleagues into the decision-making process. Because when people have a say, they identify more strongly with the project. This is how performance and action go hand in hand.

With all due respect to teamwork, which is confirmed in almost every manager magazine, here are also a few critical remarks. Not all tasks are suited to being agreed upon by a group. Constant discussion can slow down the decision-making process, significantly. It can be fatal, especially when quick action is required. Every team requires a structure. Without some measure of hierarchy, nothing works. There has to be someone who has an overview of everything that's going on and can give his veto when in doubt. In the same way that there is a trainer and team captain in the world of sports, there also has to be a general manager holding the scepter in his hand in a company. Many people tend to envy their boss. He has it so easy, he can just delegate the work. But that is a gross misjudgment. Being the boss does not just mean making decisions and bossing other people around. If you are the leader of a company, you have to juggle a lot of things in your thoughts and carry the full responsibility when something fails. A boss has to be able to handle the pressure.

A CEO must be able to set the pulse for the further development of a project and must not get lost in the details of daily work. He must produce new ideas, chew them over and see the potential that can be gained from them. Studies have shown that creativity in groups is hindered and self-criticism is too easily repressed. That shows: It's not just about the team, but about the team leadership, the CEO. Continually improving the team, increasing cooperation are absolutely worthy goals. But this does not happen on its own, it requires intelligent leadership and sometimes the strong hand of the boss.

At Karatbars, I have managed to build a team of 530,000 affiliate partners, who market our products around the world in close contact with customers. It would be utopian to believe that I knew every individual partner. However, all of these people are connected by a joint vision. And this vision is embodied by Karatbars. And although I know that it is illusory to get to know all these people personally, I still try to have as close of contact as possible. Because with all the advantages that digitalization brings, it is still personal conversations that support good team work the best. One personal talk often helps more than hundreds of emails. We notice the other person, get a picture of what they are like and gain understanding of each other. A feeling of community develops that goes beyond cultural and language differences. This feeling stands for long lasting success.

3.
"FOLLOW YOUR DREAMS
AND DON'T LET ANYTHING
GET IN YOUR WAY."

The path to success is not straight. Even when you have a good idea and you can advance it with passion, you will still face obstacles. In order to achieve your goal, you need a clear and well-thought-out strategy. But even the best laid plans cannot take into consideration all the events to come in the future. An economic crisis, the loss of important employees or difficulties in your private life are seldom predictable. These potential consequences are hard to plan for. Accepting this fact is the first step to a productive approach to doubt. We cannot predict everything. Our lives are too full of diversity and randomness along the way. But with the right attitude you can succeed in steering your life assertively in the right direction.

I still remember the shock that hit me and my family when my grandmother was diagnosed with Leukemia. This shock was followed by a painful process. My grandmother gradually lost the ability to move about. One could see how the disease was affecting her life. She eventually became completely bedridden. We found the decision so difficult, but in the end we had no other choice and organized a place in a home. A few months later my grandmother died and I sank into an extremely deep period of grief. She had accompanied me through my entire life, had raised me and cared for me. She was always present and had been a huge part of my life. A difficult phase. But there was also another side. Because it soon became clear that my mother now got on a lot better with my stepfather than ever before. There was no longer a third party to throw in their two cents or try to influence my mother. An unbelievable harmony returned in our somewhat shattered family life. My grandmother had influenced my mother her entire life. I sensed that my mother was now able to stand more firmly on her own two feet, she became more self-confident. These experiences showed me: even terrible things, crises and shocks can sometimes have a positive effect, to set new things in motion, create new developments in us and our environment that we had stopped hoping for long ago.

In these moments, one cannot allow themselves to stop functioning, instead it's important to make clear decisions. This is exactly what the daily job of an entrepreneur is. Just the decision to found a new company is a stage that is underestimated. Money has to be invested and a lot of time will be sacrificed in this new project. Your social circles will have to get used to your new ambition. How will friends and family react when you are often stressed in the early days? Especially in the early phase of setting up a company, there are many challenges waiting that require a closer look at yourself and the people that you are working with. It's all the more important to invest all your energy into your project at this time. That you will feel doubt is all too natural, you should just not allow it to gain the upper hand and influence your behavior too much. Because one thing is clear. Being an entrepreneur is by definition full of risks. No company boss can get around having to take these risks, even when they have reasonable doubts. It's about finding the right balance between worry and self-confidence. Although unnecessary doubts can result in a variety of negative consequences for your company's development, there can also be some positive effects. For example, they can help you not to become overconfident and take on more than you can handle. But setting aside all the risk and all your passion for the project for a moment, at the end of the day, no one benefits if the boss gets burn-out. In the worst case scenario, the business has to start anew. That takes time and energy.

Doubts help make us more realistic in terms of what we can handle workwise in the future. That doesn't necessarily mean having to give up ambitious plans. Doubts remind us that everything doesn't have to happen at once. Success is only sustainable when the work can unfold in continual tiny steps.

It's important to recognize one's own weaknesses. No one is perfect. Everyone makes mistakes sometimes. Learning from this and anticipating dangers that could arise in the future, is a valuable competence that one should not overlook. Doubts are a natural and really the best way to correctly estimate one's own abilities and check future scenarios from a more pessimistic perspective. For each decision, a company should weigh up the advantages and disadvantages along with the risks and potentials. Reasonable doubts belong on a scale of decision-making. A scale is a precise measuring instrument. That's why the doubts that are weighed there should be defined precisely. Diffuse fears are not measurable. Only concrete worries can be examined against the probability of them occurring. This is why it is important to formulate your concerns as precisely as possible. Regardless of whether you write them on a piece of paper or talk openly with friends and colleagues. This is the only way possible to recognize exactly where your worries are coming from and whether they are valid. A positive side effect is that seemingly huge problems at the start can be reduced significantly. The problematic core of a negative apprehension will most often become more easy to deal with this way. Consider the mountain climber about to climb Mount Everest. Naturally, he will have some doubts about whether he will be able to achieve his goal when he first looks up at the eight-thousander. But when he asks himself exactly what concrete problems he could face, the accomplishment of his task becomes more realistic. If he is afraid of the cold at the top, he can procure the appropriate gear. If he has inordinate respect for the many days of loneliness or fears having an accident, he can put a team together that will go the distance with him. A productive approach to handling doubts is possible in almost every case. Doubts encourage using the imagination, by requiring solutions. They stimulate your own creativity. As an entrepreneur you are productive. That is not only true of your own business ideas, it's also true for how you deal with emotions. Give your doubts a productive power as well, by using them to minimize future risks. Is there really still a gap in the concept? Did I overestimate my abilities regarding a particular task and could an expert help here? Doubts are often a helpful guide in how to tackle future problems. However, it is at least as important to return to a healthy awareness of your strengths. Even when the doubts in a situation are particularly serious, one should not turn on oneself, immediately. It is better to think about how the challenges will be successfully mastered.

Those who master their handicraft find ways to overcome the emerging obstacles. Falling back to a 0 to 1 score, doesn't mean that the game is over. Losing a game doesn't mean that you won't be getting the trophy at the end of the season. Keeping your goal and your own strengths in view, even when faced with criticism or self-doubt, helps you to stay on track and advance forward. Not rushing around, not reactionary chaos, but the right focus leads to achieving your goal. Follow your passion and trust in yourself and your ability to persuade. This is superior to any doubt in the end.

Many people tend to see the negative in a situation. They look for the metaphorical fly in the soup. But instead of being finicky and searching for the tiniest of shortfalls, your gaze should always be directed towards the big picture. Keep in mind how far you've come already. Remember the potential your business ideas have in store for you and don't forget all the trust that you may have already received from others. A strong team at your back can provide you with valuable confirmation that you are on the right path. Focusing on the successes you have had so far gives you strength for future ventures. It is the source of your self-confidence that you will need to take on new projects and develop yourself further.

Faith can move mountains. I experienced this in my own family. My mother, for example, with her alcohol dependency, and constantly experiencing phases of complete lethargy, finally found the courage to get some help. That was a huge step for her back then, because alcohol had become practically the most important part of her life. But she believed in herself, believed that she could renounce alcohol entirely. And it worked. And how! After six months of detox, she was finally free. She never touched a drop of alcohol again. She even joined the blue cross, a Christian society of former alcoholics now on the wagon. She held onto her faith in the truest sense of the word and went to church every day and believed in God with her whole heart. That helped her to find the strength in herself and not in alcohol any longer.

No one is immune to crises, whether in the family or in a company. People are different, have various opinions and backgrounds, and grow up differently. Conflicts are part of life. Sometimes, it feels like you are stumbling from one crisis to the next. But let's be more precise. Seldom do difficult situations have the same cause. History doesn't repeat itself and the present is controllable. An open approach to conflicts helps problematic situations be solved more quickly. In the best case, conflicts can improve character and enable you to handle the next problem more confidently.

If you work in a team, you should always have an ear open to the sensitivities of others. Friends and colleagues can point out a new perspective to a certain problem. This can lead to new ideas of how to handle a problem objectively. All these are ways to turn negative occurrences from blockages into engines that drive the company forward.

I also had to fight doubt in the early days of Karatbars. My family feared that my project could fail. I had to invest money which was, naturally, associated with a certain risk. Would I be able to win over customers? Would the theory that I had thought out, function in practice? The best answer was—in spite of all the doubts—to simply try it out. A few of my friends argued that I would throw a lot of money away and should, therefore, be happy with a secure and permanent salaried job in a company. But that went against my nature. I had my goals clearly before me and was aware that I had to take a risk in order to go forward. A "carry-on regardless" attitude would have meant stagnation. And stagnation was the opposite of what I wanted to achieve.

But the doubts left their mark. Of course I did a lot of thinking and questioned my plans. But I didn't let the doubts take over. Quite the opposite, I was more motivated than ever to fight for what I believed in. I began to talk with my friends and family, conveyed my reasons to them of why I wanted to hold on to Karatbars.

The need to offer customers a secure investment, the variety of concepts that trading with gold made possible, the technological advances—all of this gave me the feeling that I was on the right track. I also noticed that a lot of the criticisms of the doubters touched on a general feeling of insecurity and at their core were not that convincing. The "pros" of my business concept outweighed the "cons." That validated my decision to me.

To come to this conclusion, required a few weeks of brooding and emotional ups and downs. At the end of this phase I had learned a few things. First: Doubts are part of any risk you take. If you take a risk, you cannot avoid having to think over some negative scenarios. But—second—it is exactly this doubt that is an opportunity to check your own ideas, thoroughly. Doubts give you a chance to set aside any gaps in your concept and to question the validity. If the doubts come from the outside, from family members or friends, it is worth talking things over with these people. Because they only want things to go well for you, in the end, and to spare you from making huge losses. Convincing them that the risk you want to take is really worth it, is a big task. Because the ability to persuade is a core competency of an entrepreneur. Whether with potential investors, with customers or other employees—you need to be able to back up your ideas with good arguments. Don't take it personally when your friends talk critically about your ideas, instead use the opportunity to work on your argumentation. Your friends are providing you with a stage where you can try things out, so that when it really comes down to it, you are prepared. Being open to criticism, listening and accepting criticism expands your own horizons and offers a valuable new perspective. In addition, it is an excellent chance to practice, to enable a productive exchange of ideas with employees and customers. Because your customers sometimes also want to rid themselves of their concerns, independently of concrete recommendations for solutions. Listening attentively shows your counterpart that you are taking them seriously—a basic prerequisite for mutual trust. A further strategy that I have developed in my approach to doubt, lies in the highlighting of optimization potentials. Doubts usually feed on thoughts about what could go wrong in the future. But what about simply taking a look at the present? Your business idea did not come out of nowhere: you discovered a deficit, something that disturbed you, that you wanted to improve with your idea. More concretely: the here and now is not optimal, it needs to be changed and your idea will bring this important change. Persevering with the status quo is not a solution.

The doubts that come are, therefore, not a call to do nothing, rather the stimulus to do something that you are already doing, but better. It can help to make concrete lists of tasks that have a particular priority and, therefore, are the most relevant to do first. Doubts can be an opportunity to reorganize. Possible obstacles and negative reactions can be considered from the beginning and made more concrete. Then you are ready in the worst case scenario if an imagined negative situation actually arises. You then know what needs to be done and can proceed carefully and avoid unnecessary chaos.

If someone confronts you with their doubts, consider what the alternative to your plan could be. You have chosen a path that you are convinced about. You have developed an idea, that you are certain will bring the necessary improvements. Then don't let anything get in your way. In fact—not pursuing your plans, falling into passivity, is often more dangerous than taking the next step, even if it seems risky at first. Usually, you have already invested time, energy and money in the project. The principle decisions have already been made and for good reason. So one should not let oneself be dissuaded so fast from one's goal.

The fact that you have already invested some resources shows that your project has an added value that is worth working for.

"But how do I react to a doubter in a personal conversation?" you will surely ask. My tried and true suggestion is: Show how urgent it is for your idea to continue to be put into practice. Give examples of the problems that you have recognized and explain what your company can do to change it. You should also make your counterpart in the conversation aware of the dangers that exist if nothing is done and you would, contrary to expectations, abandon your project. Your counterpart should have become aware of how important your concern is at this point, at the latest.

Here, it is important to remain objective. Do you have an anecdote to share where you have noticed the problems that you want to tackle now? Have you already received positive feedback on your business plan from other people? All this shows the doubter that you are persistently and conscientiously working towards your goal, and that in this way you are able to assess risks and more strongly overcome crises.

Just as there are always doubters, there are also admirers of an idea. In phases of pessimism it can be of great value to make contact with them or to remember their convictions. What do the "admirers" appreciate about your idea? Why do they think your project will be successful? Each coin has two sides. Not losing sight of the bright, positive side even in difficult times gives you strength when it is especially necessary.

If negative impressions in your surroundings begin to take the upper hand, emphasize the positive side of the project. By turning the coin to the shiny side in times of personal crisis, you change your perspective to one of optimism that has the effect of increasing productivity If you succeed in focusing on the fascinating part of the mission, the future prognosis usually becomes much more positive.

If, for example, employees have doubts, one can create arguments along the lines of the advantages that a decision will also have for colleagues personally. This might be a higher salary, but it can also be new, exciting tasks or simply the prospect of promotion. But these promises should not just be given for that reason. Caution and a sense of proportion are required here. Potential benefits are also linked to increased employee commitment. They will need it, especially if there are bottlenecks or the company stands before its next big step. New benefits go hand in hand with increased expectations, which should also be formulated. In an ideal situation you win over the doubters among your employees and move forward as a collective.

Doubts are okay as long as you don't begin to despair. At the same time, criticism is also good and important, as long as it does not become destructive. While the critics tend to be more objective than a potential customer who is driven by diffuse doubts and uncertainties. This makes dealing with critics easier on the one hand, but more complicated on the other.

A critic usually argues with concrete arguments, provides facts and a firm opinion. In this way, he sets up the means for dealing constructively with his criticism. Usually, critics don't come from within the family, but from within a profession. They are experts in the field in which you want to establish your company. In my case, the biggest critics came from within the financial and assets sector. Traditional and experienced experts in their fields, who often believed they already knew all the subtleties of their profession. Their self-confidence and self-security initially made an impression. But one should not let oneself be unsettled by this.

Because you are the visionary. What distinguishes you from your critics is that you are able to dare to do something and create something truly new. Your critics, on the other hand, remain an appendage in the background. Don't show any false modesty, even if the tone gets a bit sharper. You should stay true to yourself and try to win with your own arguments. Because it's the customer, not your critics, who ultimately decides on your success or failure.

As convincing as the critics may be, they are not perfect either. There is no one who has been spoon-fed wisdom. Even a critic is fallible, even if it doesn't appear to be the case at first glance.

Just like when dealing with doubts, it is important to draw the positive from the negative. It helps to categorize the criticism that has been levelled at your project. Which points really made their mark? Which concerns were well-founded and understandable?

Use the comments of the experts by adjusting your strategy in some places. Adapting your concept later is not a sign of weakness, but of entrepreneurial flexibility. The point is to put the project on the market in the best possible form in order to achieve maximum success. Objective criticism can be an important trigger for this.

I also went through phases in which I strongly doubted myself and my abilities. Not only in the run-up to the founding of Karatbars, when many things were still uncertain and it was still written in the stars how things would work out. Even after my two shock experiences with the disappearing Swiss investment company, the subsequent court case and the death of my grandmother, I went through moments in which I absolutely questioned myself and my work. But what alternative was there to simply carrying on? Somehow it had to go on. And the fact that one sometimes hits rock bottom in life, both privately and professionally, does not rule out the possibility of getting up again and tackling something new. It is difficult to start all over again from scratch. My earnings had been eaten up. I had paid off the required taxes and lived life in a state of flux—always in the belief that everything would carry on as before. A big mistake as it turned out.

But I had something more than just nothing. I had an idea. My path led me back to the financial sector. Because I still had a contract with the Stuttgarter Insurance company. And I had learned a lot in my career. Of all of those things, multilevel marketing seemed particularly promising to me. I had the idea of selling insurance policies in a clubhouse. So I rented a small house, furnished it comfortably and called it "Club Holiday." I also developed a new commission model. All the people who took out an insurance policy with me for the future received a recommendation commission when they brought in new people. I remembered the statement that the US oil tycoon and billionaire, Jean Paul Getty, had made: "I'd rather earn 1 per cent from the work of a hundred other people than 100 per cent from my own labor." That was my motto.

But before I could start the project, I had to talk to my branch manager at the Stuttgarter Insurance company. I explained my project to him in great detail, but I don't think he understood it correctly. He said just one thing to me: "You know Mr. Seiz—when the sea is calm, the ducklings can swim easily." Whatever, I had gotten my permission to start. I was ready to go. And how! Within a very short time I had 100 people in the business. I had great sales, more than three times that of many of the branch managers in Stuttgart combined. My store manager was delighted with my success. But he also became a little suspicious. To get an impression of my work, he visited our clubhouse "Holiday." He checked to see if everything was okay, but he still hadn't fully understood. When the insurance turnover rose to over 30,000,000 DM per month, I received an invitation to meet with our Stuttgart branch director. In the meantime, three quarters of a year had gone by and I was back on the road to success. My income amounted to over 25,000 DM per month. However, I already suspected what the gentlemen from the Stuttgarter Insurance company wanted to ask me when he invited me to his home. "Mr Seiz, how do you do that?" the director asked me. I explained everything to him, down to the last detail, just as I had explained to my branch manager, before. He just nodded. Then he said he would soon send me his decision on how he wanted to carry on. About three months later it came. The bad news. The Stuttgarter Insurance company forbade me to pursue my work as it was. I had to put an end to the business. Once again I had to give up something that I had successfully built up and in which I had invested a lot of time and energy. Once again I had to build up something new. But that wasn't the worst part. This feeling was familiar. The problem for me was the impression that neither the branch manager nor the director had really understood my strategy.

In my career, an affirmed critic was often more helpful to me than any doubter, no matter how determined. After all, what distinguishes a well-founded criticism from diffuse doubts? Doubt is dominated by an emotion, a gut feeling of uncertainty. But there is inevitably a quantum particle of risk in any prospect of success. That's how it works.

A solid criticism, on the other hand, is certainly rational. It testifies to the fact that someone in the field has considered your plans very deeply. Criticism is a sincere form of support, if you will. It draws attention to potential weaknesses in the concept. Makes you aware of where you can still improve things. Some people will call in a consultant for this. If a critic comes up to you on his own, you can only gain from this—in knowledge, expertise and experience.

You can accept the criticism, turn a few screws and improve your own concept. If the criticism doesn't make sense from your point of view, nobody will force you to deal with it for long. This keeps the time lost within manageable limits.

Of course, there are also different qualities of criticism. A superficially formulated opinion should, therefore, be left in the lurch as quickly as possible. As well-intentioned as an opinion about your project may be, it costs you time and energy to deal with it. One should, therefore, prioritize which aspects of the criticism are relevant. Put your focus on those aspects of the critique that really sound convincing to you.

There are certainly some black sheep among the critics as well. Pedants and whiners who don't bring up their arguments out of a professional interest, but want to show off their knowledge or put obstacles in your way. Maybe they are jealous of your creativity, maybe they want to nab those customers themselves, with a similar concept. Unfortunately, this happens over and over again. Especially, when you're beginning it can be difficult to figure out who really wants to help and who wants to do harm. With experience, however, you will be able to separate the serious criticism from the noise.

Therefore, be open to criticism, but only as long as you don't allow it to shed doubt on your core convictions. You are the driver of your project. Criticism is a tool for fine-tuning, but never a reason to leave the car half-finished. Just as when dealing with doubters, one should also maintain a holistic view when dealing with criticism. What a critic finds problematic, might be loved by a customer or employee. Criticism and praise are two sides of the same coin. One cannot exist without the other. Their task is to ensure an optimal balance, with the overall aim of advancing your company.

That's what it's all about, after all. Taking criticism and doubt into consideration is not what leads to success. What counts is performance. Actions are more convincing than words. People aren't going to remember what you once said. But they will measure you by your performance, by what you have achieved.

Here, I would like to put an end to a widespread misunderstanding. Many people believe that performance automatically leads to success. But that is not the case. Performance doesn't equal success. An athlete can have the best performance of his career, run his best time, and still not get the gold. Nevertheless, dedication and 100 per cent willingness to give it all are basic requirements when creating your vision and being able to achieve your goal. Just as an athlete needs years of training to achieve their best performance and qualify for a big competition, an entrepreneur has to work on himself every day to reach his full potential.

To get the best results, the performance has to be just right. But success does not only depend on performance. There may be times when you feel that you have already reached your limit. The success you wish for may not occur immediately, without this resulting from something you've done.

If you consider a football or relay race team, for example, this quickly becomes obvious. A player's performance may increase the chances of success for the entire team, but it is no guarantee of victory. Even if everyone is functioning at 100 per cent, success still cannot be certain. Because the competition might be just as strong or a brief, random moment could have a decisive effect.

Equipped with this knowledge, it is easier to deal with failures. A career rarely advances upwards in a straight, steep line without detours. There will be ups and downs, challenges and difficulties, but also solutions. If you ever fail to reach a goal you have set yourself, this is not necessarily because of your own performance. In these difficult moments, it is much more important to have the certainty that you have done your best.

It was exactly the same in my case. With my "Holiday" clubhouse and my new commission model, I had created a successful concept that was outperforming all the previous insurance sales in the region. Nevertheless, they put a stop to my work. That's hard to endure, but it couldn't be helped, back then. I had done my best, invested a lot of time and energy. The ban from the director of my branch made it impossible for me to continue reaping the fruits of my labor. But the next thing was on its way. I already had the future in sight.

From the 1st of January, 1998, the German Telekom company lost its monopoly status. A good friend told me that a new telecommunications company was setting up a new organizational structure in Germany. The basis for it would be multilevel marketing. My specialist field, my supreme discipline. "Wow," I thought: passive income—we'll earn the money just by one person calling another. This was my chance! We were only talking in nickel and dime amounts, but they could add up. The potential was huge.

So I called all my business partners together and explained the situation to them. I was totally honest: "The Stuttgarter Insurance company has banned our current model. But I have a new plan for us." I made our brand-new opportunity to earn money on every phone call clear to them. What used to be Telekom's exclusive business, was now going to be profitable to us. The feedback was overwhelming, everyone was thrilled. And almost everyone who had previously worked with me at the Stuttgarter Insurance company joined the new project.

The company was still in the pre-launch phase. There was a lot to do and, therefore, many possibilities of how to set it up. I took it upon myself to establish a Germany-wide structure. My contact at the telecommunications company was a real professional in network marketing. I followed him every step of the way, from city to city. We stayed in hotels all over Germany. Almost every evening we had conversations with business partners he already knew. And the result was fantastic: 95 per cent of the partners wanted in. The project was something special and we were not the only ones who saw it that way. In the beginning, I was just sitting in the meetings, watching and listening. But I thought to myself, I can't build an organization passively. I had to become active. So the next day I set off for the city, full of energy, and spoke directly to people.

On the first day alone, I had approached 30 people and tried to convince them to come to our meeting in the evening. I thought at least ten to 15 people would come. In fact, the result was much more modest. Three people came. I learned quickly. About 10 per cent would come to the meetings in the evening if I had done my job well beforehand. So we carried on like this, from city to city, and I gradually built up my organization. Every ten to 14 days we returned to the same hotel. And slowly things got going. Because the partners that had been recruited brought new business partners with them and so it continued. Once it had got going, the pre-launch phase became a great success for me. I had built up the second largest organization in this company.

My first check was for 8000 DM. All this just from one person calling another, and the tiny contributions added up to the massive sum. There was nothing better than a large organization to make the model lucrative. After one year, the company carried out a big introductory tour through eight German cities and we got at least 1000 guests in each event location. From then on it really took off. We now were earning a lot of money. We travelled from city to city, arranging meetings and got more and more partners on board. It was a dream come true!

I had reached my peak. A great success—often easier said than done. But I had developed some helpful strategies out of my previous experience.

There is a reason why I talked so much about my idols Elvis Presley and Martin Luther King at the beginning of this book. I didn't just mention them because their actions were a great motivation to me at the beginning of my career and in my youth, but because they helped me to continuously grow. Not only do I admire these two Americans, I also use them as a benchmark of my own performance. To me, Elvis and Martin Luther King are examples of people with a great, personal presence, whether on stage or at the podium. They stand for conviction and ambition, courage and charisma. Both embody the attributes that have been decisive for my own achievements. They are the checkpoints I use for aligning my actions and evaluating my own performance.

Measuring oneself against great role models can mean setting oneself up for a big fall. Not everyone can become a star. Not all people are able to set things in motion that advance us as a society. The space in the history books is limited. But despite all this, our idols provide proof of what can be achieved in a lifetime. They show what is possible with persistent performance and going the last mile. Elvis was also once an unknown little boy from an American suburb like many other thousands like him. Martin Luther King also had to fight early on with the repressions that society imposed on him. And both of them managed to build something really great, something that will be remembered for decades, perhaps even centuries to come. They used their abilities to the best effect and were successful in their own way.

I have learned that I don't want to just imitate everything that distinguished my role models. Each person has his or her own individual path in life, has a different background, grows up in a different place and in a different time. But our role models are excellent teachers when it comes to improving oneself. They show what qualities and abilities are important. Their path in life is both maxim and orientation for our own lives. Because although everyone has to go his own way, surprisingly there are still many parallels between people. For example, when it comes to the ability to persuade. Just as I have to convince an important customer in a conversation, Elvis had to be able to inspire his viewers. Just as Martin Luther had to summon up the courage to talk openly about the problems in the USA, I took a risk with Karatbars. The way our role models deal with crises is an excellent guide for our own actions. It is often very revealing to examine the demands they made on themselves and compare it to our own. How did an Elvis Presley convince his audience? Which stylistic devices, which language did Martin Luther King use in his speeches? These are all starting points for improving one's own performance. You may not be able to copy an Elvis Presley—many people lack the voice to do so—but you will find a very individual way to make the symbolic value of his work fruitful for your own personal ambitions. Even stars don't have only positive sides. That's why it's not possible to make a one to one copy. Bring the positive side of your idols to the fore. That is your yardstick. Your motivation. Your goal.

Admittedly, it is not that easy to immediately fulfill all of your own high expectations 100 per cent of the time. That wouldn't necessarily be helpful either. Because it is usually the unfulfilled desire that makes you keep striving to improve yourself. Because it is the driving force for further development. Achieving a certain goal is usually only an intermediate step. New challenges quickly come into view. A competitive athlete is rarely satisfied with one medal. He wants to keep performing at top level. And the competition isn't sleeping. New competitors will enter the scene. Then new ideas, better strategies and more flexibility are required. Even if you have mastered one important stage, you should not rest on your laurels.

But you often don't get the chance. Because nasty surprises and rapid change can never be ruled out in the business world. This was also true in my case with the telecommunications company. I had been extremely successful for more than a year and had built up a profitable organizational structure. And then the bomb dropped! The company filed for insolvency. I couldn't understand it at all. After all, we were making huge sales. But again I had no alternative but to accept the here and now. My hands were tied. Later it turned out that the company had only collected the customers and then sold the contracts on at a high price. Everything had been a big fake. But what could we do? I was only a small cog in the wheel. Too insignificant to denounce the deception, too weak to make up for the mistakes and grievances of others.

From my experience as a manager, I know that everyone has a blind spot. Some lack quality in their personal exchanges, others lack a continuous focus or entrepreneurial foresight. Use the new challenges to tackle your weaknesses, constructively. And this is where the doubters come into play again. They can point out where you still have potential for improvement while the project is being further developed. Being able to accept criticism is a skill that some people have to work on.

Even when success confirms one's own performance level, one should not stop working on oneself. After I was able to report the first major economic successes at Karatbars and the company began to grow faster and faster, many came to congratulate me on these results. Of course, I was very grateful for this and I would not deny a certain satisfaction. From a sales representative in wealth management, I became a businessman of distinction. The ceremonial admission to the Hall Of Fame of Manager Magazine was a clear sign to me of this personal leap in development. In addition to the big names in the German business world, I now also found my own listed. Great appreciation for my work, but no reason for me to pack up the tents and leave Karatbars on its own.

This recognition was to me the longed-for validation of my work. The positive response to the first steps in fulfilling my vision. The proof that I had taken the right path and invested my energy in the right things.

You might think that everything became easier for me from that moment on. But that would be far from the truth. Success is not a matter of course, but the result of years of effort. And success can go away just as fast as it came.

As the saying goes, "Pride comes before a fall." I didn't want to fall into this trap. With the courage of the moment and public recognition behind me, I developed even more concepts for the further development of Karatbars. Of course, the positive feedback made a lot of things easier. But it is just as important to never lose sight of the future in these moments of success.

When praise is raining down on you from all sides, a self-critical look inside is a cure-all of preventing yourself from falling into passivity. Of course, interim success is something unbelievably constructive. But it must not obscure your view of what's essential. To me, it became the next projects I carried out for product portfolios that I wanted to further expand for Karatbars. Just as one should hold on to the positive in times of crisis to mobilize courage needed for the stony road ahead, in times of success one should focus on those aspects that still have room for improvement. The energy you will need to overcome the hurdles along the way can be drawn from these tributes. Public esteem mobilizes forces within you that should be used to propel you forward, rather than letting them settle into a permanent state of self-praise.

And that also goes for employees. As CEO and founder of Karatbars, I am often the focus when it comes to receiving recognition. Considering the risk I took in founding the company and the resources I put into developing the ideas, it's understandable. But Karatbars, like many other companies, is not just the result of a single person's performance. Behind the success there is a strong team of employees and investors. When I was honored in 2015 with the title of Senator awarded by the German Federal Association for Economic Development, it was immediately clear to me that the award was based not only on my own performance but also on the performance of all my colleagues and supporters. It's possible to come up with great ideas, such as that of an honest currency using gold, on your own. They cannot be implemented on their own. Finding people who work with the same passion every day to realize this vision, who are committed to the success of the company and who accept occasional overtime because they identify with the company, is of immense importance along the way. The recognition suggests that I started Karatbars on my own. What's true is that I probably contributed the greatest part to it. But without the many contributions, without the teamwork of the various experts, such a success would not have been possible despite my great commitment.

"United we stand, divided we fall": "We" is bigger than "I." This applies even more, the greater the challenges become. A holistic and global corporate strategy, as we pursue at Karatbars, can only succeed with a powerful team. This also includes our more than 420,000 business partners worldwide. They work every day in 120 countries around the world to implement the Karatbars concept. Karatbars is no longer just Harald Seiz. Karatbars has become a community. The goal of making a better life possible for mankind is no longer the single mission of a lone wolf, but the project of an entire global community. The many awards I have received as the initiator of this movement are a great acknowledgement. But they are by no means the only reason for the pride and joy that I experience in my day-to-day business with our customers and partners at Karatbars. Their respect and trust is the best reward and thanks for the great work I have put into Karatbars.

This anecdote shows more clearly than any other example I could give, that it is not just a question of one's own performance. To be successful, you need a team with team spirit. A team that may have small frictions, but that always puts the common goal in the foreground. Even in seemingly individual sports, such as tennis, winning the title is never only due to the performance of the individual player. Behind a Boris Becker you'll find a whole team of coaches, psychologists, doctors and, last but not least, friends and family members. It is only through their contribution, through their support, that the professional's talent comes into full bloom.

4.

"EARN THE TRUST
OF YOUR CUSTOMERS
EVERY SINGLE DAY."

4 20,000 business partners—that's about the size of a city. But unlike in a city, there is no anonymity in the network of customers and partners of Karatbars. There are many reasons for this, but I would like to highlight one very important one because it is so essential for a successful business. The customer is the "bread and butter" of every company. No demand, no business. The recipe is that simple. That's why it's so important to do everything necessary to win over your customers. And not just at the beginning, it must be continuous. How can this be achieved?

The search for answers to this question provides an entire portfolio of options for achieving this. Of course, the product that has been developed must provide added value for the customer. Of course, you have to make sure that people become aware of you, that you are recognized in the market. These are undoubtedly important factors in being able to reach new customers in the first place. But if you dig a little deeper, you can uncover the decisive aspect of a good customer relationship. It's about trust. I've recognized this in every new project I've started.

Shortly after the failure of the telecommunications company, I was already pursuing a new project. I knew how new beginnings felt and already had a promising idea. Back then, eBay was a big deal. My idea was to implement eBay's concept in network marketing. I was absolutely convinced that we would grow quickly with this idea and maybe even outpace eBay in the end. But first we needed a well-functioning online platform. I had set aside a few euros and hired a programmer. I designed my own network marketing plan and gathered together a core team of 20 people around me: my executives. Even before the programming was completed, we started setting up the organization. Our service offering consisted of packages for €400, €800 and €1200. Many of our customers started with the big packages. Week after week I carried out presentations. Mostly in Aalen, Germany. Because most of our business partners came from there. However, I quickly noticed that our approach was eating up a lot of money. In order to build up the business to the size we wanted, we had to organize elaborate events, invest a lot of time and also rent the appropriate rooms. Programming was also a cost factor. So we needed more funds to really be able to invest properly. "The whole undertaking stands or falls with that," I thought. In my search for solutions, potential sponsors came into my focus. I talked to some potential financiers. And some were quite taken with the project and showed a willingness to support it. In all, the sponsors agreed to cover investments of €250,000. In the same breath, however, they made it clear that the whole thing could still take some time. After all, these sums still had to be approved by company management. And they also wanted to wait and see how our young project would develop. The problem was, we still had our running costs. In particular, I had to pay the programmer who was crucial to our work. Because without a functioning and attractive platform, we lacked the necessary means to show our sponsors the potential of our project. As a result, the payments from the financiers were sluggish. And our own revenues were not yet able to cover the necessary financing at this early stage. So I kept waiting for the investors. After one year the programming was as good as finished. And finally I was able to present something solid.

We had created a platform similar to eBay, just a bit better. We offered additional options, such as extending the selling time for bids in the final minute or concealing the winner of the highest bid. I worked very hard on setting up a new organization. Slowly, but surely, more and more people joined in. We now had over 700 business partners. But I was still waiting for the green light from the investors. 25 per cent of the company for €250,000—that was the deal. At least on paper. The months passed and nothing happened. We were able to keep the project going, but there were still no big profits to report. Money even got tight, from time to time. Once, when we couldn't pay the programmer, he blocked the site. A catastrophe. Our partners became insecure, our customers angry. The programmer had sent us a shot across the bow. To me, his behavior was incomprehensible. Of course, he wanted to see his well-deserved payment. But nobody was denying him that. We were all in the same boat. We all wanted to go forward and create something. To me, we were a team and not just a group of egotistical lone wolves. We had a common goal, or at least that's what I had thought. But I had to learn that a programmer thinks differently.

Nevertheless, I still believe this: A human being is a social being. Although we can get by on our own for a while, sooner or later we will seek contact with friends, colleagues or family. Since the beginning of time we have been in contact with each other. In our early childhood we learn to communicate with each other. Even a newborn baby will expect an emotional reaction from its parents. Our sense of social need is a primordial human instinct.

The motives behind it may be very different. One person strives for friendship, the next for productive cooperation, another hopes to find the love of his life. But all these motives are connected by the desire for social exchange.

A life in isolation is a nightmare for the great majority of people. We need each other to be mirrors of ourselves. We need interaction in order to understand ourselves better. We need someone who will catch us when we feel bad, give us an honest opinion when we are stuck in a conflict. The list goes on ...

The contexts in which we long for another person are manifold. It's all the more astonishing that in the business world people like to and quickly tend to think of the customer as a mere economic unit. Thereby, forgetting the social diversity. In the statistical reports of large companies, "the customer" appears as a number on a sheet of paper. I don't want to deny the usefulness of statistical surveys and analyses. But my aim is to draw attention to what is behind the façade, behind the economic facts and figures. People are the focus.

Every customer is also a human being at the same time. And in the modern world, everyone is at the same time a customer. You don't have to be conscious of this all the time, but every day you take care of a multitude of business transactions. Whether buying coffee in the morning, filling up your tank on the way to work or when moving into a new flat. The customer is not an abstract figure blurred on the horizon. He is a person with very concrete needs, desires and behaviors. He is complex, shaped by friends and family, has his own opinion and his own personality. What does this insight mean to the CEO?

Against the background of the complexity of people described above, we have to learn to think from the customer's point of view. This means accepting the diversity of customer needs so that you can make an offer that is optimally tailored to their individual needs. Do not try to see the customer as an obscure mass phenomenon, instead try to imagine yourself in his personal point of view. Thinking from the customer's point of view will help you to recognize the potential for your market activities and make them more fruitful later on.

As a company boss, it is all too easy to lose contact with ordinary people. You sit in your ivory tower, sign documents and move further and further away from the normal life that takes place every day in the reality at the bottom. That is fatal! It may seem wonderful and fill you with pride to be enthroned in the executive suite, only giving your signatures. But in order not to lose your innovative powers and above all your contact with reality, constant contact with the customer is your "alpha and omega."

Because you don't want to develop anything in the end that completely ignores people's needs. Don't want to produce anything that devours resources without generating demand or ends up in the garbage can of history as quickly as it was once brought to life. It is precisely these negative experiences that I would like to spare you from. Because these things hinder the further development of your company. They inhibit economic progress without which you will not achieve your goals. Don't put up new obstacles in your way. Poorly-thought-out service offerings cost time, effort and money and have no positive effect. To avoid that: Keep your feet on the ground, no matter how big the company becomes. And keep your sights on the customer, and the customer's point of view.

Thinking from the customer's point of view means keeping an eye on the market you want to serve. It is a huge fallacy to believe that you are the only player in a certain market segment. At the very least, when attentive observers recognize that your idea is successful, they will try to come up with similar, even better concepts. In an open market economy, anything is possible. Friends can become competitors. There is no guarantee of your piece of the pie. The customer, who is looking for new service offerings, will also look at what your competition has to offer and make an assessment of who he is going to do business with. You should do the same. By comparing the product range of your competitors with your own, you gain important insights to better position yourself in the market.

But this is not to say that you have to surf along immediately with every new wave that hits the market. A new trend can quickly become obsolete and leave a few long faces behind, especially those of the entrepreneurs who have speculated prematurely and without consideration. What you need is an awareness of which service offerings meet customer needs accurately over the long term and over time.

Who doesn't know the story of the toddler who—once the new toy has been bought—immediately throws it in the corner and leaves it lying there, forgotten. The fascination for the new on its own is not enough to be successful in the long run. Especially in the fast-moving times in which we live today: what is the hot potato today is off the shelves again, tomorrow. Only long-lasting, quality products and services ensure continuous growth.

I thought I had created exactly such a promising product for the future with our platform. But everything turned out quite differently. At the end of 2002 I had a meeting with our investors. Out of nowhere, they cancelled their investment. It was a shock! And it got even worse: Uncle Sam came knocking at the door. Suddenly I had the public prosecutor's office in my house. The reason: progressive customer advertising. Hardly, did I become aware of it and the officials were standing in our office, taking everything with them that wasn't nailed down. Invoices, contracts, documents—all of our files went into a criminal police van. There I was, looking at an empty filing cabinet, desks that had been ransacked and a bunch of dejected employees. Was that all I had left? After all the effort, after all this hard work?

At that moment I fell into a deep hole. I didn't understand what was going on and how the prosecution had come up with their accusations in the first place. I always wanted only the best for my company and our customers. Honesty was my top priority. Why should I be considered a criminal?

I was invited to an appointment with the criminal investigation department. The officer who sat across from me in the interrogation couldn't stand me. I felt that immediately. His piercing, rigid gaze, his eyes squeezed together, he tried to put me in my place. The whole atmosphere, the police station, the men in uniform and the rough tone caused me a great deal of anxiety. I just wanted to leave. I wanted to get out of there as quickly as possible.

I was so unappealing to the police officer because he had the completely wrong idea about me. He probably thought: "I earn my €4000, while this shady businessman makes millions compared to me." I noticed his anger under the surface. Anger arising from envy I couldn't understand. He had probably completely misjudged the amount of effort I had put in and the risk I had taken to be successful. And my project had not even been crowned, yet, with success. I was further away than ever from my first million. People like this criminal police officer begrudge you everything. No matter what you do.

I was totally open with him in the conversation. I had nothing to hide, after all I had always been honest in my opinion and so I explained our business model to him. He then said: "Such a thing is forbidden. You could even go to jail for that." That sentence struck me like a bomb. Now it was there: the great fear. Panic swept through me. What if he was right?

The next few months were complete torture. The worst part was the uncertainty. I had no idea what to expect. No information from the department, absolutely none. And then, finally, the trial came and I could get clarity about my future. With my lawyer at my side I felt a little more secure. And when I saw the judge, I felt even better. He had very positive charisma. But charisma was neither here nor there. Primarily, he stood before me as a guardian of the law, as the supreme authority whose decision I had to submit to. That was his job. To him, the facts mattered above all. And they were in the material that the public prosecutor's office had confiscated from us. Next, the police officer was given the floor. His statements came across extremely negatively, even downright condescending towards me, in parts. Unfortunately, the bad impression he had made on me during the interrogation was confirmed.

Fortunately, the judge also noticed this and soon made it clear that he would handle the matter with neutrality. The proceedings ended with a ban on my business model and a fine. No prison! My worst nightmare had been averted. A heavy weight fell from my heart. The relief was magnificent. But, unfortunately, it didn't last long. Because only a short time later I felt the melancholy pressing down on me like lead. My business lay on the ground in shatters. How could it continue now? But even though I didn't see much that was positive in that moment, I knew that there would be a light at the end of the tunnel someday. I had to stay stable, trust in my abilities. Then the success would come again, even if the way there would again demand all my perseverance and patience.

Because success is always for the long-term. Quality products last for years, if not decades, on the market. A well-founded consideration of which market developments you will follow and when you will stick to your course, is required here. Because even rearranging a company also consumes resources, which only pays off if you have made a strategically far-sighted decision.

However, meeting the essential customer needs is only part of the work you have to carry out. Add to this the correct communication of the advantages of a product. The target market plays a decisive role here. Who do you want to reach? And above all, how will you reach your target group? The usual types of mass media, from the press to television, have long since only represented a small part of the spectrum of possibilities available to you. Here, it is also worthwhile to think digitally. More than 80 per cent of Germans now use the Internet. The trend is rising. For a modern company this means that it has to make use of this space, itself. The spectrum ranges from personal newsletters and advertisements to social media marketing. The success of campaigns is soon measurable because of the technical possibilities. Advantages that should not be dismissed by a wave of the hand and should be made use of for your project.

The opportunities of reaching potential customers are more diverse than ever. In order to design the optimal approach, we at Karatbars follow a two-step process that I would also like to share with you. Roughly speaking, the work begins with an open brainstorming session in which all possibilities of communication are put down on a list. This is a method that quickly and creatively shows all the communication channels without placing too many restrictions on them yet. Only in a second phase, are individual approaches evaluated according to their feasibility, effectiveness and expected impact. With this type of filter mechanism, the best ideas arise, so that you are able to reach the ears of the customers.

In this way, a few basic points are described, which ensure that you get off on the right foot with the customer. But now to the basis of every successful customer relationship: Trust. How can you build trust with the customer?

I personally maintain close contact with our customers all over the world. Even though it has become impossible for me to establish personal contact with each individual, due to the sheer mass and range of interested parties, I still try to make mutual exchange as open and individual as possible.

As the CEO of Karatbars, I have an important role model function for all of the employees and partners. I can only be credible if I act according to my own philosophy. Only then will our partners be aware of our standards and be able to implement them in their daily contact with customers.

By establishing a dense network of cooperations all over the world, we try to reach our customers locally. In each country you will find different circumstances and possibilities. It is, therefore, the task of our consultants to adapt the central message and service offering of Karatbars to the local situation so that they can unleash the optimal benefit for our customers.

The mentality of customers can also differ substantially. It makes a big difference whether you are giving advice to a customer in Germany or in South Africa. Having local people who are aware of these differences and can take this into account in the consultation is worth its weight in gold.

Karatbars stands for financial stability, security and trust. We want to carry this concept out into the entire world. Because this concept is universal. Everyone has the need for security and being well-positioned to handle crises. We need local experts so that we can achieve this as well as possible everywhere.

I wrote about expertise in Chapter 2. Expertise develops over time and evolves with the challenges. For our global network of advisors, this means that our client advisors must feel comfortable in what they do, locally. Only then is there a prospect of long-term cooperation, in which the quality of the work increases over time. On one hand, because a professional communication routine exists between the head office and the local office. On the other, because every client advisor continues to improve his skills with every sales meeting, every new offer, every new negotiation. Long and stable business relationships with our partners are a top priority. Because continuity creates expertise and trust, not just between the business partners. As a rule, the client also wants the advisor to whom he has once built up trust to continue to be there for him. All sides profit from a constant cooperation: the company, the business partners and the customer.

If you want to address a specific target group with your project, you should also keep their perspective in mind when selecting consultants. Depending on the exact target group you want to reach, you should pay attention to the specialization of potential consultants in the field. An insurance expert will not always be the best financial advisor. An affiliate with an old customer base will not necessarily have a soft spot for digital products and, therefore, does not provide optimal access to young, technology-savvy new customers. Furthermore, a previously and specifically trained consultant brings with him important qualities for you and the customer.

Firstly, he will be able to demonstrate more expertise thanks to his previous knowledge. He can present tailor-made solutions to the customer and knows, in the best case, about the peculiarities of a consultation. This is an art in itself and must be mastered. How do I use my body language? How do I find out what is particularly important to the customer? How do I create a familiar atmosphere in the conversation? These are only a few of the questions that can be answered correctly or incorrectly in a consultation and that contribute significantly to success or failure, to the conclusion or rejection of a contract.

Secondly, an expert will be more able to provide you with well-founded feedback than a newcomer. This will allow you to improve what you offer and adapt it to the respective foreign or domestic market. Thirdly, a consultant with access to your target group offers great potential to help your business grow. In the end, the same applies here: All sides benefit if your consultants are equipped with the specific required competence for their target groups and if you manage to hire them: You, the business partner and the customer.

But it's not just the personal contact to the customer and competent advice that counts. At least as important is the fact that what you offer is just what's needed. Your product must be designed in such a way that it is always useful for as large a number of people

as possible and is, therefore, in demand. In an ever faster changing world, you need to be flexible and be able to adapt to survive
in the marketplace. In other words, the work within the company
must also include innovative components. The development and
expansion of new business models is essential in order to stay in
touch with the pulse of the times and the needs of the customer.
A partner who is a good customer advisor can provide important
impulses and introduce you to new concepts that you can make
profitable, in the best case scenario.

The world is constantly changing. This means that your product is always being given a different meaning. In order to position
it correctly, it is necessary to understand major changes going on
in the market. That in turn, is being impacted by societal change.
Despite all individuality—customers are still a part of this society
that shapes them. And that strongly influences what they buy.
Many products are simply about presenting something to the outside world. Status symbols such as jewelry, cars or real estate ful-
fil important functions, and promise social acceptance. What the
customer wants, therefore, always depends on how his environment values certain objects. If your product fulfils a higher social
purpose in addition to its standard function, its sales will continue
to increase. Societal change is also reflected in people's economic
needs and in political changes.

It was exactly these observations that led up to the foundation
of Karatbars. I realized that almost all currencies are mere promises made by the country issuing them. They do not represent a
real value, but a fictitious sum, an insurance one does not know
for certain will be enough in times of crisis. This development
has manifested itself with the termination of the Bretton Woods
Agreement and the dollar's decoupling from the gold standard.
The debt crises in Argentina and Greece have confirmed this.
In addition, the financial crisis in 2007 made it clear what dramatic effects risky speculations could have. Entire asset portfolios
became worthless. Formerly, highly profitable investment securities fell into a bottomless pit. A landslide, a jolt, a shock for countless savers.

The safe investment, because it is crisis-proof, has become fashionable again. Recognizing these changes, I looked for suitable solutions. And I found them in the gold trade. The first product I developed along these lines was "Cashgold." A means of payment in the form of a conventional card. A card with a small but subtle difference. Because in it was embedded 0.1g of fine gold, which represents the actual value of this new currency. This protects "Cashgold" from state currency devaluations and financial crises. Because the Goldcard only promises what it actually and truly contains: the value of the embedded gold, which aims to be one of the most stable on the market. That's why I like to call "Cashgold" the most honest currency in the world. Honesty—that's exactly what I expect from each of our partners. More than anything else, the client appreciates this virtue. And it is an expression of my innermost conviction and the core concern of Karatbars: an honest currency through honest advice. No crooked deals, no small print, no tricks. These methods are obsolete and no longer have a place when it comes to stable investments.

I would like to emphasize that the "Cashgold" business model was only the first step towards building Karatbars. If I had stopped at this point, I would not have developed Karatbars further with new business models, and the big success would never have become reality. Success comes from innovation. Germany is the best example of this. We cannot win with low wages or poor environmental standards on the world market. That should not be our goal either. Instead, the German economy thrives largely on the sometimes highly complex innovations that experts are developing, year after year. I also see Karatbars in this tradition. That's why we added new products to our portfolio at an early stage. It's all about the customer and covering the entire spectrum of his needs in the field of gold investments.

A good idea, an initially successful product is only the prelude to the establishment of a large company. Don't stay in one place after you start. Otherwise, you run the risk of being taken over by larger groups, as the saying goes: "Big fish eat little fish." And you are also putting sustainable corporate success at risk.

Because the market is in a continuous state of flux. Just take a look at some of the recent phenomena that is changing our lives in the long term, as proof.

Take urbanization, for example. More and more people are living in cities: The result is a shortage of apartments and rising rents. An innovative real estate concept would be the best solution. Architects, urban planners and real estate experts should have the answer, here. Or digital networking that makes communication and the consumption of media content possible across borders around the world. Amazon, eBay and Netflix have become market leaders. The last example of major changes in today's world is rising life expectancy.

Demographic change is on everyone's lips and solutions for a dignified life in old age is more in demand than ever. In view of the shortage of skilled workers, technical inventions from the field of robotics could be implemented to support this trend. A market of the future.

If one only considers these three developments out of the many possible, it quickly becomes evident what a dynamic field the entrepreneur is in. He must not only keep himself up-to-date with the latest knowledge, he must also be in a position to adapt his company in such a way that it can react effectively to these changes.

Reinventing oneself is not just a concept from the psychology of personality, it is equally valid for the company as an adaptable organism. Just as the creatures of the earth have adapted to changing environmental conditions in the course of evolution, companies must grow in line with their environment. Innovation and the courage to put the new ideas into practice are indispensable growth requirements for every company.

In this sense, Karatbars has also developed new products in addition to the Cashgold system. We have developed KaratPay, for example, in response to the digital future. An online platform that can be used to check your account balance, make transfers and pay at selected shops. All in one app, mobile and in your pocket on your smartphone. We have not closed our minds to the trend of digital currencies either and have, therefore, created our own cryptocurrency: the KBC. With these novelties in the product range, Karatbars is well prepared for the near future as well. However, this does not mean that no further ideas will follow. We are constantly working on new opportunities to do with the concept of a gold-based currency. The strategy of constantly promoting innovation has more than proven its worth. It has made Karatbars what it is today: a globally operating company with turnover in the millions.

New products that are geared towards new living situations not only ensure that existing customers remain loyal to your company. Ideally, you can also reach new customer groups with high performance innovations. Digital services, for example, are essential for younger customer groups. And your existing customers will probably also expect you to move with the times. You can be a first test market in which new products can be offered and improved in practice. They provide you with valuable feedback and will view it as special appreciation when value is placed on their assessment. This assumes you have new business ideas that you can test. But how does innovation succeed? What climate is needed for creativity and a creative spirit?

Creativity is a fleeting phenomenon. Many people find it difficult to be creative on demand. A flash of genius cannot be planned. Just as the hunter waits in the bush for a deer to pass by, the entrepreneur may also have to be patient for ingenious ideas to arrive.

We can discipline ourselves to master routine tasks, but creating something really special at the push of a button is not how things work. You need the right moment. You can't create this moment out of nowhere, but you can make some decisions that promote innovation and clear your mind for inspiration. Not only for you, but also for the entire team: a positive, open attitude is water for the mills of an innovation-friendly business climate.

Frustration and fear are the biggest obstacles on the path to productivity and inventiveness. These emotions are part of every person's personality—as well as joy, sadness and humor. They are all right and good, they make us human and give life its enriching nuances. But in an entrepreneurial environment they prevent sustainable success.

Frustration also shaped my emotional condition after I was banned from doing business by the court. There were also problems in my private life. The doctors had diagnosed cancer two years earlier in my mother. She had had to undergo several chemotherapies in order to maintain a chance of survival. Then things got a little better, but nobody could say for sure whether the cancer had been defeated or not. We had to live with this uncertainty.

My hands were tied by the court ban. And these professional and private incidents drained me of a lot of strength. Strength that I now lacked to tackle something new. So I stayed at home for the time being. I lacked the drive, the drive to pick myself up again. My thoughts only circled around.

But at least I wasn't alone. I met my girlfriend in 2001. In the beginning I thought: "This is my big love." I was head over heels in love with her and felt intoxicated when I was at her side. Unfortunately, this feeling did not last forever. Increasingly, we clashed, sensed more and more the things that separated us. Love means growing together, supporting each other and helping each other. And it was just this quality that was too seldom the case with us. We had an on-off relationship. We argued a lot. Not infrequently about money. She worked in the management of a large company, had a regular and secure income. I, on the other hand, was dependent on the success of my business ideas and the very start of my projects were often marked by financial uncertainty. To support her, I also took care of her two sons.

But at some point I felt the walls closing in. Day in, day out looking at the same four walls, always the same thoughts in my head, which weren't getting me anywhere—my situation wasn't getting any better. I couldn't sit around at home any longer. I needed new impulses. The situation was unbearable. I had reached the lowest point of my life so far. I had to start something new. But how could I avoid failing again? How could I be spared another K.O.?

At first glance, the realization I gained, here, was anything but encouraging. Because the possibility of failure can never be completely ruled out in entrepreneurship. But the good news is that we can learn to deal with fear and frustration. We can focus our minds on what we want to achieve and thus spur our minds on to perform at their best. But what can we do to create a productive atmosphere? What are the paths to more creativity?

In order to avoid frustration at an early stage, the first thing to do is to adopt the right attitude. Of course, you are ambitious and want to advance your project forward as quickly as possible. There is nothing wrong with that. Because it is precisely this ambition that helps you get the best out of yourself. But it can also turn into frustration if you take it too far. What do I mean by that?

Creativity takes time. There's no use putting yourself or someone else under constant pressure to be creative. Often, this only blocks the flow of work and distracts from actual projects. If you want to be productive, you have to allow yourself to take some breaks. That doesn't necessarily mean staring into the abyss and twiddling your thumbs. Breaks can be creative. Either to complete other tasks that are already clearly defined or to look for new inspiration. Successful people are open to the diversity in the world. They absorb impressions from all areas of life and later make use of their new ideas in their work. You can get important impulses from creative breaks that enrich your work on a particular project. Even a genius like Albert Einstein was not constantly occupied with physics and mathematical formulas. He was also politically active and took part in a social life outside his professional world. Art and music can provide important inspiration for thinking outside the box. I myself found an important balance to my professional activity as CEO in music. My Elvis performances make me feel free. The impressions I gain inspire me and give me strength for my work as the head of Karatbars. Encounters with people, moments on stage, experiencing myself singing and dancing with my heart and soul—these are things that can be transported into my professional life. These moments are sources of new ideas. They are ways to rethink the world. And being creative means nothing more than seeing things from a new perspective, conceiving things in a new way.

My career start as an investment advisor.

My own financial company—our christmas party in 1988.

My first sports car.

Fast cars were and remain my passion.

Headquarter of Karatbars in Stuttgart.

One of my numerous moments of reflection—always with my vision in mind.

A historic hour for me in 2018: issue of bonus coins. It's the first time in the world that people can exchange their digital crypto coins for real, physical gold.

At the trade fair Invest in Stuttgart.

At one of our events in Canada.

My visit to the Freedom Celebration Event in Los Angeles 2016.

At the Token Fest, a crypto fair in San Francisco in 2018.

A wonderful morning before Token Fest in San Francisco in 2018.

At the New York Stock Exchange in 2018.

In a TV-Interview in front of the trading floor at the New York Stock Exchange in 2018.

At the Federal Association für Economic Development.

At a fair in Doha, the main city of Katar.

Harald Seiz on the set.

Newly printed: our Cashgold.

Genuine 24-carat gold is used in our Cashgold production.

Phases where creativity is lacking are inevitable in the long run. Accepting this is important to help avoid frustration later on. One should, instead, see the phases of creative calm as an opportunity to absorb new impressions and broaden one's horizons. In today's business world, a chronic lack of time has become the standard of everyday quarrels. The art, however, is in consciously making more time. Even for things that do not seem to immediately lead to a concrete goal. Time to try things out, experiment and let yourself be inspired can unleash incredible creative potential. By withdrawing once in a while from the hectic of everyday life, taking a step back and reflecting on the world, a true miracle can come into existence. To be creative, you need to distance yourself from time to time from everyday life. You need a clear head that allows you to see the big picture. It is not a good idea to spend your life running around in the hamster wheel, spinning it round until the bitter end. Get out, at least for a short time, and reinvent your wheel. A productive time will come, again, and the more inspiration you take from your rest phases, the more productive your time will be.

Ideally, you should plan time for these rest periods and integrate them into your work schedule. In my case, most of these breaks arose by chance, due to breaks in my career, but looking back, I now view them positively. Because I have always tried to make something positive out of my misery. I did not hesitate for long to jump back into the game, for example, after my online marketplace idea failed. All in all, I had gained some experience in networking-marketing, and was able to quickly and efficiently develop a new company structure. That's just how it turned out. I gathered together a team of 50 people and we started selling insurance policies. It worked well, very well. Soon we reached a turnover of €10,000,000. But that was still not enough for some of my employees. This wave of success spurred them on to want more, and so they dared to go their own, new ways. They detached themselves from joint projects in order to earn even more money on their own. They simply took the completed contracts with them, and in the end, despite all my work, I found myself no longer swimming in money but wallowing in debt.

The debt burden threatened to crush me. I needed an idea with which I could earn money quickly and without being dependent on others. I needed a flash of genius.

You know the old saying, "necessity is the mother of invention." And luckily, that was the case with me in September of 2004. I was sitting at home in front of the TV zapping through different channels. It was looking to be one of those miserably long days of complete boredom. Until I came to a show about smoke detectors. Something pretty banal, I thought at first. But then it started to dawn on me that smoke detectors are very important because they can save human lives and prevent major damage in an emergency. But who had a smoke detector at home at that time?

That was my salvation. I decided to sell the smoke detectors from door to door, directly to the customer. I bought a small stock of smoke detectors at a price of €5 each. I offered the devices for €25 including a warranty. I put on a warning vest and was ready to go. My office was in the business premises of a friend in the middle of the main shopping street in Stuttgart, Germany. I had everything I needed. And that's how I started out. I was independent, self-reliant and free. With a great sigh of relief, I went out into the hustle and bustle of the city center. I was convinced that the smoke detectors offered a real value to everyone. And I believed that people would quickly recognize that. And so it was! I managed to sell about 40 a day. That added up to a lot of money. But the money did not go into my pocket. Because, in the meantime, I had accrued €250,000 in debts with my other projects. To really be able to start again, I first had to pay them back. That had the highest priority.

Things were slowly going uphill again, at least professionally. My relationship was still in a bad state. My relationship to my girlfriend had become only stress and more problems. At the same time I couldn't live without her. But I eventually had to face reality. Even though my girlfriend had grown very fond of me over the years, I gradually realized: "If something is over, you just have to let it go." But it's not easy. Especially, when feelings are involved.

So I sold smoke detectors for four years. I threw myself into my work partly to distract myself from the difficulties in my personal life that surrounded me. Besides, I was on the lookout for new networking-marketing models that I might like. But none of them really convinced me. I had become much more skeptical. I now had a lot of experience in the financial sector and I knew that many of the products were not good enough for the customer. Much of it was just smoke and mirrors and a little hot air. One prime example: life insurance. When you see how life insurance policies are handled by insurers, it makes you feel sick. A German court made it permissible to call life insurances "legal fraud" in one court case. That basically says it all. In the vast majority of cases, the insurance company may invest 50 per cent of the investment sum in such insurance policies for its own purposes. What security does the customer have if it all goes wrong? What happens when the money is suddenly gone? The customer is the one who suffers in the end. To me, this is fraud. A business shouldn't be allowed to work like that. The customer's needs must come first.

Nevertheless, insurances are still supported by the state. I did not want to take part in this game. That is why I was looking for something that is honest, that works and that can benefit people. I was creative and had made a success of my idea of selling smoke detectors. But that shouldn't be the end. I sensed an urge in me to continue developing new ideas. I only had to give my creativity space so that it could turn into something productive as soon as possible. But as the saying goes: "Good things take time."

When applied to your team, this means that you should also make sure that your employees have some free time at their disposal. As important as a structured workflow may be, sometimes it is the deviation from the daily routine that can set new and productive stimuli in motion that positively affect the innovative power of your company.

Once you have overcome frustration or at least reduced it, you should take on the second big enemy of creativity. I am talking about fear. Fear is one of our most primordial emotional states. Fear is a kind of early warning mechanism to protect ourselves from an expected threat. Our ancestors had to decide whether to react to their fear and flee in the face of a threat or use their aggression to take up the fight. The fight-or-flight response. So fear is a central emotion for us which helps us react to external dangers. But we should examine these threats in two ways.

Firstly, in terms of the probability that they will actually occur. Many people tend to get bogged down in negative scenarios that may seem possible, theoretically, but are extremely unrealistic in practice. With this negative distortion of the future, it is all the more difficult to be courageous, to dare something great and to leave room for creativity. Because it takes courage to give free rein to one's ideas, to put oneself in uncertain situations in order to try something new. What others may think about it is of secondary importance. It all depends on you and the daring you invest in giving your creative "I" the air it needs to breathe.

Secondly, one should ask oneself what exactly one is afraid of. We often carry vague fears around with us. Like little sandbags they burden us, robbing us of our lightness of being, making us sluggish and inflexible. Usually, without good reason. If there really is a good reason for our personal fears, we should talk about it. And at the same time describe the fears as precisely as possible, restricting them and learning to understand them. This is the only way to master your fears. If one closes oneself off and remains alone with one's fears, it could bring on an emotional cycle of worry. This has gotten to the point that scientists are now talking about the phenomenon of the fear of fear itself. In order not to become trapped in this vicious circle in the first place, you should always try to understand your concerns in the most concrete terms possible. Talking with other people about your worries helps you see the circumstances in a new light and to differentiate real dangers from merely imagined, mostly far-fetched disaster scenarios.

Even if a bit of doubt remains, this is no reason not to be courageous. You need to be able to face the challenge sometimes, even against your own better judgement. For some fears have a performance-enhancing effect. By simply exposing ourselves to seemingly dangerous situations, our body reacts with positive tension. Here I mean, the infamous "adrenalin kick." This helps us to achieve a completely new level of performance. Who hasn't experienced the fear of exams? The fear of failing, of being left empty-handed in the end? And yet many people master exam situations extraordinarily well and in some cases even significantly better than when preparing. Having confidence in yourself and your abilities is crucial. Like a professional athlete, you can outdo yourself in moments of the highest emotional tension. You regularly see the "underdog" triumphing, unexpectedly. This scenario was epitomized in Greek mythology when David defeated the seemingly invincible Goliath in battle.

It's the same with art. Think of the musician who has stage fright shortly before his performance. Would he stop the show because of it? Of course not, he's used to the feeling and finds a professional way to deal with it. Even more to the point—he knows how to translate this feeling into productive energy. For some, this feeling is a great incentive to keep returning to the stage, again and again. I also learned to appreciate these moments a lot. You learn to overcome your fears. You feel that it's worth being courageous, to move forward, not to hide. Others will admire you for your positive attitude and you will discover a new facet of yourself. What could be better?

As your experience grows, you will notice that fear decreases as a factor in your business decisions. Because every decision inevitably involves the risk of making mistakes. Mistakes are as much a part of business as the "amen" in church.

But a mistake does not mean the end of the world. No company has gone bankrupt because of a single mistake. Errors can even play an important role. Firstly, because we can learn from mistakes and secondly, because we no longer need to be afraid of mistakes. Everyone makes mistakes, life goes on anyway. This realization seems to be easier said than done. But it is of enormous value for a creative attitude. The courage to make mistakes once in a while, to accept little failures, will add new dimensions to your ideas. You have to think in new ways to be successful. New paths always entail uncertainties and risks. Which decisions were wrong or right, one frequently only knows later. It's easy to be wise after the event. Therefore—don't let yourself be sidetracked by those who come saying: "I told you so." Talk is cheap, action is what it's all about.

Some of the greatest inventions were initially dismissed as aberrations, belittled and considered to have no future. Mostly by those who focused only on their own world and not on the future. For example, a renowned professor at Oxford's elite English university is said to have described Edison's model of a light bulb as an "obvious failure." What an error of judgement! At the latest, by the end of the world exhibition in Paris in 1880, the light bulb idea would be forgotten, or so went the speculation. There is hardly a better example of misjudging the enormous potential of an invention than this case. A building, a room without a light bulb is unthinkable in most places in the world, today. The light bulb is by far the number one light source. Whether at home, or in public buildings, even street lamps are equipped with light bulbs. With his invention in 1880, Edison encountered a huge growth market.

Just think of the German's favorite pastime and you'll find another example of unfairly maligned inventions. The petrol-powered car by Carl Benz, patented in 1886, was dismissed as a mere "flash in the pan" by none other than the then German Emperor. A serious error, as it turned out later. Soon cars were to replace horse-drawn carriages. Today, they are an indispensable part of the cityscape.

These two examples show that new ideas, however absurd they may seem at first, can unfold enormous potential. Not only for the development of one's own company, but also for the advancement of mankind. The car has become an integral part of human mobility. Whether in cities or in the countryside, whether electronically or petrol-powered, whether as part of a sharing concept or for private use. The invention of the car has opened up a multitude of new design possibilities. New business ideas were able to follow, an incredible dynamic. One that only very few people could have imagined from the outset.

This leads me to a key sentence in this section: Believe in your ideas! No one can say with absolute certainty whether an invention or new business model is doomed to failure or correctly positioned for success. The range of possible scenarios is huge. Mistakes and failure are just as much a part of it as success and optimum decision-making.

Of course, mistakes can have very painful consequences. Either for you personally or for your company as a whole. That's why a well-founded risk analysis has to be part of the decision-making process. Especially in smaller companies that want to establish themselves on the market, poor decisions can have a serious and negative impact on market opportunities. In larger, already established companies, there is a tendency to take more risk because there are more resources available, if things go wrong. Depending on the situation in which your company finds itself, you should sometimes take on more risk, sometimes be more cautious. Always with the premise in mind that economic success arises when new shores are reached, new paths are created and new ideas forged. He who doesn't dare, doesn't win.

In this sense, failure is not really a defeat, or a breakdown, but an opportunity to learn something valuable. Firstly, about one's own weaknesses, secondly, about the implementation of new ideas within the company and thirdly, about reactive market mechanisms. All these insights will allow you to better implement future innovations in your specific field. In the end, you do not emerge weakened from your mistakes, but strengthened. Enriched by important experiences and knowledge that can be used in the future. In this way, an initial failure can later give you a competitive advantage over inexperienced competitors. The school of hard knocks is a tough way of getting through failures without losing faith in your future success. But anyone thinking the path to the first million will be a walk in the park is facing a sharp learning curve ahead. To build a profitable company from scratch is a laborious task and requires perseverance. But it is also an incredible experience that will shape you for the rest of your life. If you don't even try, you've already lost. Lost a valuable chance to live your life as you choose.

Being an entrepreneur in the land of my dreams, the USA, embodies exactly this thought. In this country, the amount of failed projects an entrepreneur has gone through is not viewed as an expression of incompetence, but as an expression of a great wealth of experience and a special entrepreneurial spirit. The failed entrepreneur is a person to be respected. A person who, despite failure, didn't give up but got back up again. Who stubbornly and sincerely believed in himself and his own ideas. A person who did not give up hope, even when the going got rough. In this respect, we in Europe can learn a great deal from the United States. The American dream is still alive. It can become a reality in each of us. I do not join in with the skeptics, but see the opportunities that have been given to all of us. With intellect, creativity and perseverance anyone can make it to the top. You don't have to dream, you have to make your dreams come true. Step by step, meter by meter, with the persevering goal of achieving more than you already have.

In order to draw the right conclusions from mistakes, however, you need management with the right approach. It would be fatal to stumble blindly from one trap to the next. On the way up there will be obstacles lurking in the shadows. Don't march into them blindly, despite all your courage.

Instead, you need to make sure that you actually take something with you if you have made a mistake. What has happened should be thoroughly processed, both emotionally and mentally. The reasons behind a failure could be hiding some uncomfortable truths. Honesty is the highest priority here.

It is easy to place the blame of a failure onto external circumstances or other people in order to protect yourself. This may seem reassuring at first, but in the end it won't get you any further. Only those who openly deal with their mistakes and question their own behavior will achieve the best learning effect. I call this the positive awareness of mistakes.

In the fight against the fear of making mistakes I would like to spearhead the need for a self-confident attitude towards risk. Mistakes are rarely carved in stone. As a rule, a mistake will not pursue you throughout your entire life. Time passes—what might be a painful fact for some is an advantage for the entrepreneur. He can set his sights to the future and not remain imprisoned in the eternal past. And the future can often put an error into perspective.

Because people forgive and mistakes can be corrected. The fact that you have made a mistake does not necessarily mean that this will be to your disadvantage forever. The sooner you realize that an action is going to be a mistake, the faster you can react. This significantly reduces the negative effects for you and your company. Just like major diseases, such as cancer, early diagnosis is crucial for your chances of recovery. Sometimes the cure is costly and associated with painful side effects. But let's look on the positive side: even mistakes can be "cured." With countermeasures, corrections to the course you are on and reorientation, you have a whole range of entrepreneurial options at your service. It is not a matter of trying to avoid mistakes, all the time and everywhere. To be continually successful in the long run is a utopian idea. The probability of errors is significantly reduced by attentive management, but it is almost impossible to erase them completely. That's not what it's all about. Rather it's about having the competence to deal with identified mistakes in such a way that they are, first of all, less serious and second of all, a huge gain of experience for the future. It is important to find the optimal way to correct errors in accordance with the situation at hand. Discussions about guilt and blame are useless. Only a precise analysis of what has happened allows the right conclusions to be drawn. How could the error occur in the first place? Could it happen again? What needs to be changed for the future? These are the crucial questions that you should discuss with your employees.

You should also consider the nature of the error in order to set up an appropriate response plan. Was it just a minor mistake due to carelessness? Or is there a structural imbalance in the company that would lead to further failures? Certain competencies may be lacking or the work processes may not be coordinated sufficiently. The range of potential sources of error is broad. Following these up requires meticulousness—above all and especially on the part of the management personnel.

Conflicts between team members is a classic example of when a CEO is needed. A failure can also simply be the result of an accident that is no one's fault. However, this interpretation should not be adopted too hastily over a thorough and honest root cause analysis. Too often, external circumstances are used as a pretext to avoid having to openly address real-life shortcomings. The problem is that you only push the cause of the error further into the future and don't tackle the barrier to success at its root. Like a criminal investigator looking for the perpetrator, the entrepreneur must also follow the clues to get to the problem. Otherwise, the real problem stays hidden and will lead to new difficulties, later.

If you internalize this approach, mistakes no longer have to mean a broken neck or leg, but become a challenge that you can overcome by facing it proactively. The same approach you adopt in the development of new ideas or preparing for a customer meeting, works when managing failure. Tackle it energetically, with a positive attitude to improving things in the future. Self-pity and bitterness have no place here. Pragmatism and a strong hand are required.

Failure and entrepreneurship, inevitably, go hand in hand. You will definitely have plenty of chances to see the positive in the negative, and plan more appropriately. You can prepare for dealing with mistakes. It makes sense to identify potential sources of danger and to keep possible countermeasures in mind, especially for a CEO. Fast and effective measures for the correction of errors can significantly limit damage. Sources of error can arise within the company or from external influences. For example, internal work processes can get mixed up, or employees are not able to provide their best performance for personal reasons. It is best to start with an honest analysis of yourself. Open and honest self-reflection on your own strengths and weaknesses gives you the opportunity to make your work as effective as possible. Here you can decide which tasks you can take on yourself and where you need to delegate tasks to employees. If everyone can use their strengths to the advantage of the company and their weaknesses are not forced into their daily work, the right workflow can be created. This is how a group becomes a team. The results: more output, a better performance record and a better quality of results.

I would like to give you an example of this. An experience I had with Karatbars, that confirms this point, exactly. In order to make Karatbars position stronger for the digital age and ensure the greatest benefit from technical advances, we decided to develop an app. This was easier said than done. Because an app consists of a wide variety of building blocks, each of which require different skills in its development. These skills must be intertwined to create an attractive and compelling result.

There is the technical component: programming and configuring. But there is also the esthetic component: What should the user interface look like in the end? How can the structure be designed clearly? Not least, but first and foremost, we had to consider what functions the customer would need from with an app. There were three levels that had to be worked on in parallel and in constant consultation: the technology, the esthetics and the functionality. In terms of the functions, an app offered many advantages that would make gold investment even more convenient and versatile. Because you would always have access to the app via your smartphone with Internet access. Perfect for keeping an eye on your gold investments. We developed a total of four functions that were made available via the app. Upload Gold enables you to upload new gold to your account. Physical Gold makes a payment possible from your gold credit account balance from the comfort of your own home. Buy with Gold rounds off the large growth market of cashless payment by mobile phone. And finally, Gold Back, a special feature for our partners to have their commissions credited directly to their app account. I was the driving force behind the concept for the content of the new app. But I lacked the necessary know-how to set up the entire app on my own. I, therefore, had to delegate design and technical programming. I trusted in the skills of our employees and partners. So I finally succeeded in developing an optimum end product in the form of KaratPay.

Some people may have a soft spot for the possibilities of digitalization, others have a feeling for the information technology issues, and there are those who know how to make an app appealing and user-friendly. They have all brought in their strengths and contributed to a great result. The development of the app can safely be described as an across the board success.

Because with this app we offer our customers a modern addition for managing their gold investments. We make use of the advantages of digitalization: permanent retrievability, convenient payment and the transparency of all data. For some, it is a nice addition. For new customers, this app may even have been the decisive factor in choosing Karatbars. In any case, this app has increased the benefits for our customers. The app combines transparency with flexibility—a bonus point for every customer.

A few more words, however, on external sources of error. A company is always situated in some kind of physically external environment. It profits from demand and from infrastructure, but it can also be affected negatively by external forces. The most obvious example is negative market developments that can put a squeeze on demand. Very finely tuned antennas are needed to detect problematic developments at an early stage. In this way you can gain important time to align the company to the new conditions.

The market, in turn, is heavily dependent on the interactions between politics and society. Is there a diplomatic crisis? Are trade relations on thin ice? Are military conflicts on the horizon? These are just a few of the factors that influence the market. One cannot unravel all possible scenarios to figure things out down to the smallest ramifications, especially since much remains hypothetical. But depending on the business model or entrepreneurial orientation, some factors can become particularly relevant.

However, one thing should not be forgotten in all the internal and external crisis scenarios: An entrepreneur's biggest asset is his customers. The depth of a customer's trust can survive even the most difficult of times. It carries the company through both good and bad phases of its development. It is worth investing in it. Each and every day.

5.
"NEVER REST ON YOUR LAURELS."

With this book, I have set myself the goal of showing you the path to success. You already know the most important coordinates of a successful company. Passion, competence, high performance, self-confidence, courage and creativity are the six big cornerstones of turning a great idea into an empire.

As important as these fundamental virtues are for every entrepreneur, there is a question that shouldn't be left out. What if success comes all at once? How will you deal with fame and wealth?

To ensure that the rapid flight of fame is not followed by an equally rapid crash, you must be able to answer these questions with exact precision. It may sound paradoxical. But you also have to learn how to deal with success. There is a widespread misunderstanding that those who become successful can lean back and put their feet up. No, there is simply no guarantee of success for all eternity. But the way you handle your success can make a significant contribution to ensuring that a golden present is followed by a golden future. If you turn the right screws and pay attention to the right things, the dream of lasting happiness does not have to remain a dream.

My success story began at the end of 2008. That's when I heard for the first time about the small gold bars that could be used as an investment or security. My interest was piqued, immediately. I did some research on the Internet, browsed through all the pages

I could and devoured several books. I read up about the dollar as a reserve currency, its various modes of operation and its origins. I was shocked! The stories ranged from conspiracy theories to the Illuminati. I thought to myself, "Enough is enough!" I didn't want to have anything to do with it. That's when I stopped the research on the dollar. The most important thing had already become clear to me: Gold is a stable and safe investment. Finally, I had a starting point again, a hint of an idea where I could begin a new project.

Things also changed dramatically in my personal life. The relationship with my girlfriend was, by now, over. At least that's what my mind was telling me. My heart, however, was still attached to her. Nevertheless, I knew that our relationship had no future. I continued to take on responsibility and watched her children from time to time in the evenings so that she could work. My girlfriend told me that I could stay, but the last thing I wanted was to fall back in to the old dynamics. Because sooner or later we would have been unhappy again. That's why I gradually began to distance myself and that was a good thing.

The separation was a great challenge for me and was very draining, on the one hand. But on the other hand, I also drew new strength from it that I would need in the future. I took on a whole new inner attitude when I saw the film "The Secrets" and listened to the audiobook of the "Master Key System." A revelation! From now on I focused on what I wanted. And not on what I didn't want. I looked ahead. I also began to visualize my goals. €50,000—I wanted to earn so much money continuously. So I hung a fictitious banknote over my bed. A €50 bill, to which I added three more zeros. Every morning and evening I looked at it and gained enormous drive. I had a concrete goal and I was ready to give everything for it—each and every day.

Now I started working with gold. And the good thing was: I could fall back on my network of long-term partnerships. Partners who had remained loyal to me over the years and my various projects, because I had always been honest with them. A company from Munich sold the small gold bars. I made contact and the CEO

came to my office. On that occasion I invited my twelve best partners to come in as new executives. The CEO had us completely
persuaded, down to the last detail. He spoke with unparalleled passion about what gold really is. He absolutely radiated with passion
when he answered our questions. It was thrilling. He showed us
how the governments of the world use paper money to spend currencies that are becoming less and less valuable due to inflation.
He showed us how many currencies have been devalued in the
last 100 years to the point of total depreciation. Nothing more than
debt currencies, I thought, and that's how I saw the light. Gold
as a stable and crisis-proof investment, regardless of what happened in politics, regardless of the power games going on between
the global players—that was the answer. And my answer to the
CEO's fiery speech was: "We're in. We are going down this road
together." From that point on, things only went uphill. I didn't
think it was possible. But within seven months I had built up an
organization of more than 20,000 people. And in the seventh
month, I was already earning €100,000 per month, twice what I
had planned for. I had exceeded my own expectations. Thanks to
my many years of experience, the trusting partnerships and not
least an ingenious business model. Everything fit together. It was
fantastic!

For many, huge success is something unusual. Everyone
experiences highs and lows in the course of their lives. This also
includes the small moments of success in everyday life. A positive
business deal, a passed exam, a profitable assignment—all this
makes us happy for a moment. But these experiences are nothing
compared to a really big coup. Huge success breaks over you like
a wave. A wave of happiness and joy, but also a wave that washes
all kinds of new things ashore. A new situation arises.

Those who are successful may not be able to grasp their happiness at first and may act imprudently in exuberance. With success
comes public attention. Suddenly being the center of attention can
be an unfamiliar feeling, which you don't know how to deal with.
I would like to show you how you can deal with these new experiences. I also know from my own experience that you can save

yourself a lot of inconvenience if you know how to behave before-hand, when everything seems to be going great and the sources of money seem to be inexhaustible.

My first tip is to stay calm. Strength lies in peacefulness. Of course euphoria is an incredibly great feeling and those who are successful deserve to enjoy these great moments. Take a moment to enjoy. That is the reward for years of hard work, for overcoming inner and external resistance, for the personal and entrepreneurial development you have undergone. That is your right!

Nevertheless—after the first exuberant moments, after the ecstasy—life goes on. For you personally, success may have changed a lot, but the world around you has not been affected. With inner peace you can learn to understand your situation better and to deal with it in such a way that success does not remain a one-time event, but becomes a constant companion in your life.

The first thing you should try to do is to keep this great feeling of happiness alive in yourself. Euphoria and pride are positive sources of strength and energy. Especially in difficult times, which can never completely be avoided, the memories of these emotions will help you to develop stamina, so that you can also walk the difficult pathways. Inner confidence gives you courage and self-confidence—success factors par excellence. Without a good helping of this, I would probably never have taken the risk of an entrepreneur's life. I still remember well the times when I was a vacuum cleaner salesman for the German Vorwerk company. Being paid on a commission basis—the more business I did, the higher my salary. This demanded a great deal of commitment from me, but also gave me the freedom to find my own way of working. From then on, I had to approach strangers on my own, convince them to buy our products. In the beginning it was unusual and I was nervous. Would people reject me? Could I even earn enough money with it? But soon I found my own work rhythm and learned to appreciate the interpersonal contact. I had entered a new situation. And I was more than satisfied with how it went. Sometimes you may not have the courage to make a big change, although a fresh impulse is urgently needed. That's exactly the time that memories of great successes can help. They let us feel that we can make something great out of anything. They remind us that you will also find your way around in new situations. Memories of success let you see the great potential, because you can draw confidence in your own abilities from them.

The first success can be the spark needed to set off a great career, because you begin to sense that you can do this and anything is possible. In order to retain these special moments, you can come up with your own, personal symbol that is closely linked to your memory of success. What it is, exactly, whether a special piece of jewelry, a photo or a bank statement of the company account, doesn't matter as long as the thing has a personal value for you. Sometimes a thing from the outside world is necessary to help reactivate inner forces. The innumerable sacred buildings around the world are a good example of this. By admiring the beauty and grandeur of these buildings, we draw strength. We marvel at the achievements man is capable of. A level of achievement that you can also bring to your own work.

It cannot be said enough that man is a social being above all. Which is why recognition from peers and society also plays an important role in terms of how much strength you can draw from your successes in the future. It may sound a bit superficial, but status symbols continue to be an important door opener and can lead to external validation. It can also work wonders for your own self-image. So don't hesitate to put your success on display. But, definitely stay authentic. Whether wearing a tailor-made suit, a high-quality Rolex, or driving in your new sports car—if you stay centered, you will radiate positive energy. Who doesn't want to be next to a well-dressed person? Who doesn't want to take a spin in a high-performance sports car? Some people may think that talking about status symbols is banal, but they do induce a certain fascination in us and our fellow human beings. Let your enthusiasm show and others will be happy to be enthused. This is also a part of the good life.

Demonstrative status symbols are not everything, of course. Once you have hit the jackpot, you will not be admired for your wealth and fame alone, but for what you have achieved through your own efforts: your creativity, the courage to take risks, the clever rationality that you have demonstrated to get this far. "The journey is the goal" and a successful person has usually displayed these qualities on this journey. First and foremost, these virtues are what distinguishes you and why you are being admired.

What counts is not just the material, the bank account balance, the fancy car or the expensive watch, there's also a personal component. Success is more than a big, fat balance on your bank statement and also more than the new sports car. And I'm saying that, although I'm a big car lover. But if success is more than an expression of wealth and prosperity, then what else is missing?

I would go so far as to say there are people who can live a happy and fulfilling life without the big money or perennial prosperity. Success is always very subjective. For some people success is just being able to help someone else without getting a lot back in return. Just think of the countless volunteers all over Germany who are doing great things without being paid. My sister, with whom I still have a special relationship, is a geriatric nurse. As you have probably heard in the media, you don't earn big money that way. A nursing shortage and poor working conditions—these are the buzzwords associated with this kind of work. Of course, she also suffered from it. But what matters most to her is that she can be of help to other people. That makes her happy. Everyone she was able to care for well, each instance of invaluable help— that is success. Money is not a huge factor here. What she receives is incredible gratitude. An immaterial value that cannot be compensated by money.

There is also potential for personal happiness in the small things in life. It all depends on one's perspective. The ambitious businessman will of course think on a larger scale and not be satisfied until he achieves his vision. But the small, intermediate steps leading to the quantum leap offer reason to be proud.

Your inner satisfaction counts more than the material status you achieve when you are successful. There's no question, you will be happy when you can afford things that were unthinkable before. Status symbols, a great vacation, gifts for friends and family—you shouldn't necessarily go without life's great amenities. They make life more comfortable. But they shouldn't obscure your view of what's essential, namely your inner fulfillment. Success also means being at peace with yourself—because you have carried out what you wanted to achieve. Because a thought, a once vague project, has become reality. Thanks to you and your service. From my point of view, it is this performance that will really give you fulfillment. Wealth and outward signs of prosperity are only, albeit extremely positive, side effects.

Living a truly fulfilling life is sometimes harder than you think. You have probably heard the often recited story of the rich heir who, despite all his possessions, never became happy in life. He could afford to buy whatever he wanted, but something was always missing. Something that money couldn't buy, that he could only have achieved through his own efforts and striving. As a successful entrepreneur, you are in the lucky position that money did not fall into your lap—you earned it yourself—which is much better. Through your own actions you created something special for which you are now being rewarded. The money that one earns through one's own achievements has a symbolic meaning in addition to its nominal value. It is the reward for your efforts, the validation of your activity, an expression of appreciation from your customers and society as a whole. This is what success is.

With all the praises sung to the competencies of successful entrepreneurs, one thing should not be forgotten. Great success is rarely based on the performance of a single person. You may embody the company, and its extraordinary performance, as CEO. However, this does not mean that you, alone, are responsible for the success. Egocentric people tend to forget who their assistants, supporters and partners were on the road to success. Gratitude and a little humility for the actions of the entire team are in order, here. You may have defined the overall direction and held the reins in your hands, but without the numerous contributions of others, without the reliability in performance and commitment, a company cannot get very far. You might have the best ideas in the world—developing these ideas takes qualified manpower. It doesn't work otherwise.

Just take a look at your network. There will have been contacts that were particularly valuable to you. People who provided a significant push to the development of your company. Supporters, creditors, employees—the list could go on and on. Some may have grown tired of the endlessly long thank-you speeches at award ceremonies. However, they contribute to a realistic assessment of one's own and the company's performance in recent years and make it clear that sustainable success is hardly conceivable without good teamwork.

This gratitude keeps your feet on the ground and prevents you from flying away, even when you feel like you've sprouted wings. This is another one of the biggest challenges in times of success: maintaining a sober view of yourself and your environment.

If you want your success story to continue on into the future, you are dependent on your team. And it's only obvious that those who have made a contribution to the overall success of your company deserve some recognition. In order to keep your team together, both financial and personal gestures of appreciation are appropriate. Here, too, the following applies: "Money won't make you happy," but if an increase in salary is understood by the employee to be an expression of recognition for his performance, a salary increase can have a performance-enhancing effect. Your team's affinity and identification with your company increases. And that is a good thing, because future crises and bottlenecks can always occur. Should this happen, you will be more dependent than ever on loyal employees who won't hit the road when the first difficulties appear. Gratitude is an investment in a successful future, as it were. It strengthens cohesion and gives strength in times of need.

In times of success, many doors are open to you. Because there is no shortage of resources. Demand is good, the financing stable. Good conditions for trying out new things. Instead of resting on your laurels, you should seize the opportunity and use the "additional" resources that are now available to you. If things run smoothly, you can dare to set off for new shores, and take a risk. Because this solid financial foundation you can now build on is a competitive edge.

If business is going well, you will have the time and money for new innovations. As pleasant as today is in successful times, you should also keep your eye on tomorrow by investing with foresight. Not only in your most loyal employees, by paying attention to them, but also in innovation and further development. Of course, it sounds tempting to simply sell a successful company or cash in and let the business die. You've got to ask yourself what benefit your work has for society. This is a responsibility that you should not relinquish at the first opportunity. I am a great friend of reinvestment. This is how I live up to my responsibility of continuing to improve people's lives through my work. At the same time it is also a personal need of mine, to be an entrepreneur. I want to create added value because it makes me happier. And I believe many people feel the same way.

In order to keep up with the dynamics of the market, your company needs to keep up-to-date. You should, therefore, not be too frugal with the financial resources you generated in economically successful times. Ideally, further projects within the company were already being planned and can now be implemented. If the cost intensity was an obstacle before, now is the chance to take the next steps. Those who cling to the money they gained, will soon be left in the dust by current events. The best example of this is the continually rising inflation in the Eurozone. The amount of money in your account may remain the same if you do not spend anything, but the actual value of this money is continuing to decrease. Without you being able to do anything about it. People who are experiencing great financial success for the very first time are tending to hold on to their income, feverishly, in the hope that this situation will go on forever. The exact opposite is the case in building long-term financial liquidity. Never before have the opportunities been so great for realizing your visions, as in times of success. A wise reinvestment not only promises economic growth, it also adds another notch to your own experiences.

You will only be able to discover how to operate successfully in the business world, if you try out new things and implement your ideas. Or to put it another way: He who remains in his small hometown will never know what possibilities are out there in the big, wide world. He will never be able to use all his potential. The world will seem strange to him, opportunities will be limited and he will probably envy others for their ability to be open to new things and their wealth of experience. You, too, should go out into the world and take advantage of the opportunities. Especially, if you have the necessary capital at hand.

The CEO of the Munich company did not use his opportunities. He made mistakes that could not be corrected later. He used the same marketing plan he had used for his business in the financial sector. But trading gold works differently. Adjustments were necessary. I recognized this and asked my boss if I could create my own marketing plan within the company. He didn't mind. And so I began.

Of course, my individual marketing plan caused difficulties in the overall organization of the company. Because from now on his and my people were working differently. But unlike his people, my people were earning money. We were profitable. My CEO's marketing plan was unbalanced and paid out too much money to the investors. That made it attractive to the client in the beginning, but it didn't work in the long run. The very different profits ran through the company's balance sheet like an invisible thread over time. The profits were not able to cover the personnel and financial expenditures and the company had to declare insolvency, in the end.

The gold deliveries came to a halt because of the arrears and many customers turned their backs on the company and went away. To this day, my former CEO still owes some people gold. All in all, the project came to an inglorious end. A shame, because we started out together with a lot of enthusiasm.

Still, I'll never say a bad word about my old boss. After all, it was he who gave me the opportunity to work with gold. I took my first steps in this complex business together with him. He opened the door to the business for me that now generates millions in sales. He had just invested wrongly, was exploited by his partners. He didn't have his team under control, for which he could only blame himself. So he had to take the responsibility for that.

I told him I couldn't go on like that. The insolvency weighed heavily on our shoulders. My part of the work had been extremely profitable. He understood that and let me go. Now the way was clear for me to start my own company. I founded Karatbars International in November 2011. In addition to the great success that came my way, I learned a lot of new things when I founded the company. Compared to being part of a large establishment, working to set up my own company added a completely new quality to my life. This brought with it new challenges and sharpened my awareness for the right way to deal with success.

"Time is money." There is probably no other saying that is as popular in the business world. There's a reason for that. Because it sums up the classic dilemma every businessman has. If you want to earn money, you have to invest time. But time invested in one place is money lost somewhere else, because other tasks needing to be done cannot be carried out simultaneously. Professional time management is, therefore, indispensable. But back to the real topic: dealing with success. At the moment when you can celebrate financial success, the time-money rule is in your favor. And on a massive scale. Because now you have the money you need to take the time you need. Time that you can use to just be lazy. Or time in which you can initiate new projects. And that's exactly what it's all about. With the "more" time that the money you earn gives you, you are able to become more future-oriented than ever before and you do not have to painstakingly plan out your current profitability down to the last detail. Every good entrepreneur is also a visionary. But every visionary also needs time to allow his visions to unfold. Do you remember what I wrote about creativity? Don't expect a brilliant idea to simply enter your mind out of nowhere. A really great vision develops slowly, feeds on your experience, failure and endurance. All this naturally uses up your strength and takes nerves. But in successful times you have the energy to invest without worry. Without the huge fear of neglecting things somewhere else. So it's not just about reinvesting part of your income, it's also about being able to focus your personal skills on what's most important to you.

Don't underestimate your friends and family. I often hear from successful colleagues that the people closest to them give the decisive impulses for their future life. Impulses that ultimately lead to success. To take time when the time is available and not simply carry on inside the hamster wheel—this is an art that has to be learned. Far too many people simply work through their tasks, without questioning much where it will lead. A far-sighted entrepreneur asks exactly this question. He sees the new prospects, sets his sights on new goals—far away from the daily ups and downs. In times of success, you have the resources to plan for the future and uncover new potential. What are you waiting for?

Friends and family can give you strength in every situation. Good friends are there for you, in the good times and the bad. In conversation with them, you will learn to reflect more and question your own actions. This is just one of the many reasons that stable friendships and cultivated family ties are so valuable. My old school friend Wolfgang was there for me when I was constantly being teased. He also ensured that I discovered completely new abilities in myself. I still remember how he put my name in a draw one evening. My name was pulled and I performed a song on stage for the first time. People were thrilled, threw money on the stage and I was able to earn such good money until I was 28 years old. A great experience I would never have had without Wolfgang. So as not to lose important friends, you should dedicate enough time for them, if your schedule allows for it. Friendships have to be cultivated so that you can rely on them later. It takes time, but in the end it's always worth it. My experiences with Wolfgang absolutely confirm this.

There is another phenomenon that plays an important role in dealing with success. As wonderful as entrepreneurial success may feel at first, it is not everything in life. In the end, it's about the people you help with your project. On the one hand, of course, there are the customers who profit from your products. But on the other hand, it is also about your circle of friends and family, to whom you can express your gratitude. Imagining that you become rich through your own efforts, alone, may satisfy your ego. But only rarely does this really correspond to reality. Particularly in difficult phases of life, it is always the people closest to us who are there when we need them, breathe new life into us and believe in us when the customer or business partner might not. Friendship is of immeasurable value, which you should appreciate especially in the exuberance of success.

And success is all the nicer when you can share it. That sounds paradoxical at first, but it's really true. "Shared joy is double the joy." Making others happy is sometimes worth more than the best of deals you've hammered out. Giving your friends something back for what they have personally done for you in hard times guarantees inner satisfaction that is more valuable than the tenth luxury yacht in the harbor. Such behavior is a sign of being down-to-earth but also respect. And, conversely, will engender recognition from your fellow men for your actions and how you've dealt with success. Strength of character shows itself in times of crisis, but also in times of apparent abundance.

To measure success only in money would be naive and far too short-sighted. While it may seem that everyone is chasing the dollar in a capitalist system, a bank account full to the brim is no guarantee of lifelong happiness. The truth is that those who only work for the big bucks, seldom become successful. Striving for wealth and prosperity can be a passion. But there is usually much more behind it. For example, the concept of self-realization. It is a dream to be able to go one's own way and to get rewarded for it. To see one's own vision thrive and grow is a source of great satisfaction in life. To see other people dedicating themselves to your project because you are personally convinced about it, is an indescribably positive thing. In a world where money often takes center stage, we often forget what is really important. Inner conviction, a positive certainty, the ability to do something meaningful and to help—beyond the numbers and spreadsheets, this is where real success lies.

If you display these virtues and show gratitude to your supporters, you will prove that despite your success your feet have not lost contact with the ground. This is what makes you popular. Man is a social being. Everything he does is embedded in one way or another in a social environment. We expect a reaction and appreciation from those around us, a gesture of recognition or a word of attention. As different as people's professions and passions may be, everyone deserves appreciation. This starts with the seemingly small things. Think, for example, of waste management. It's precisely because there are people who are willing to take on the task of waste disposal on a daily basis that you have the opportunity to dedicate yourself to other things. In the modern world we live embedded in a massive system which depends on the division of labor and mutual support. Without roads, transport and schools, our standard of living would not be possible. The absolute certainty that the basic things of life are functioning, simply cannot be valued highly enough. It's because everyone contributes to the whole that everyone can pursue their individual passions and continue to develop their abilities. This is the basis of any form of self-realization, whether in business or your personal life.

The key word, here, is modesty. Great personalities remain modest no matter how much recognition they receive and no matter how much money comes their way. You know yourself well and know that your performance will only bear fruit if it finds fertile ground—a complete system within a functioning community. With your company, you can make a valuable contribution to the functioning of this community. You, too, deserve to be valued. But don't forget that your contribution was only possible because many other people were carrying out their tasks. These people paved the way for you so that you can now operate. So that you could develop your creativity and allow your entrepreneurial spirit to unfold.

As a global company, I understand the value of a well-functioning environment. There are countless countries in which this is not the case, and sometimes those of us in Germany take it for granted. Imagine if your place of residence was not connected to a network of roads or you had no access to the Internet. In many regions of the world, this is still a reality. And the possibilities for the people in these regions are, therefore, very limited. It is a privilege to be able to use these technical advancements, due to the fact that someone put them in place. From the community—for the community. And it is my duty to show my gratitude and my appreciation for the achievements of others, with my contribution to the community. Great success and significant progress are best achieved when we learn to work together. When we know how to use our abilities for the benefit of each other. What would a Martin Luther King have been without his loyal following throughout the USA? Without the people who believed in him and his vision? Without the fearless demonstrators who stood up in courageous protest against discrimination, even taking beatings themselves? What would an Elvis Presley be without his fans? Without the people who gave him encouragement, recognition and support? And what would Karatbars be without our business partners all over the world, whom we trust and who support us with their competence in good and bad times?

Large projects require great commitment. Not only from those who lead them. But by all those who are involved. Some people might say: "Too many cooks spoil the broth." That doesn't apply to success. Great success is only possible in a community. To become aware of this fact, is a valuable method for ensuring a down-to-earth and realistic approach to success. Because with success comes responsibility. But more on this later.

Your answer to success should, in addition to gratitude and modesty, above all be appreciation for the action of your fellow human beings. "Human dignity is inviolable." This stands in the first paragraph of German Law. And dignity is not dependent on wealth and prosperity. Everyone deserves appreciation. Even a criminal should not be denied his dignity. Also, he was not born a bad person. Far too often it is the influences and experiences within his environment that have made him a "breaker of the law." It is a great achievement in our time, to grant these people their place in the world. Do not look down on people who are not as

123

successful as you are. Be grateful for the happiness that has come to you and at least have partially made your success possible.

Successful people often have special talents and great will-power. But even these abilities do not come by chance. A talent must be encouraged. We develop willpower in our childhood and youth. We are strongly dependent on the people around us. A mother or father who gives us strength and trust in ourselves. Someone who gives you the feeling that you can do great things. I did not grow up in simple circumstances. My father left my family to return to Greece when I was three years old. Even though my mother and grandmother usually had an open ear for my worries, they had their own struggles, which I had to cope with as a child. Their alcoholism and the associated shock experiences I went through as a young boy have left their mark on me to this day. I will never throw my life away to drug use. I will never drug myself to suppress negative experiences. I want to deal with the problems openly and sincerely, however difficult they may be.

My mother was a good-hearted woman. Before the alcoholism, she worked in a factory on a piecework basis, putting every penny she earned back into our family. Despite all the problems she had, she also gave me support. She gave me validation, gave me courage and strength. She was proud when I took my first steps as an entrepreneur. So I learned early to be independent, to endure things and to look ahead.

People who trusted me and my abilities helped me to get up again, to tackle new projects, not to give up. Just like my friends, later, who supported me every time I perform and still do. I owe my great time as an entertainer above all to my old school friend Wolfgang. He gave me the courage to get up on stage and show my stuff. I discovered my voice and noticed that I could inspire people. Even if I didn't make big money from the occasional appearances in discos, weddings, company celebrations and private parties, it was a success for me all the same. I sensed what was inside of me and noticed that I could inspire people with what I was doing. Sometimes such experiences are worth more than an exorbitant fee. These experiences shape your personality. And it is through

them that I have become what I am today: the manager of a globally profitable company.

In retrospect, I would say that my twenties had an incredibly
strong influence on me. My character is based on these experiences and the success I enjoy today. It is a part of my life story. A
story full of ups and downs, successes and defeats, descents and
ascents. Not an easy life, but one in which I always knew who my
friends were, and that my mother and my sister were by my side.
The importance of which cannot be overestimated.

This quickly becomes clear when you consider how many people grow up without any emotional connection to another person,
without advocates and supporters. They are hindered in their path
to healthy self-confidence at an early age. Self-confidence, which
has to be laboriously developed later to be successful. Because the
courage to act entrepreneurially is based on self-confidence. The
courage to make difficult decisions and take responsibility for mistakes. Healthy self-confidence is a basic quality necessary for being
a successful CEO.

It's a slippery slope when you start going down the wrong
path. Perhaps the lawbreaker just got in with the wrong crowd at
the wrong time, perhaps he never learned to assert himself when
faced with people who did him harm. Maybe in difficult moments
he was lacking someone who understood him. I don't want to start
a pity party, but there are a multitude of scenarios that could lead
you down the wrong path.

Just as many as the paths to success. Chance and simply the
good fortune of having met the right people at the right time also
play a role. I still remember meeting a friend of mine in Gerlingen during my first job as a beverage supplier who told me about
the job opportunities in the financial sector. He told me about the
opportunities for advancement, about the freedom and independence that one could gain as a financial advisor. That appealed to
me. If I had not met him, I would probably never have become a
financial advisor, leaving so much of my life's potential untapped.
I wasn't paid very well for delivering beverages. I managed to make
ends meet with my 800 DM per month, but only just. Working

as an independent vacuum cleaner salesman sounded much better as I could clearly earn a lot more on a commission basis. And that's the way it was. I had more than tripled my salary within a few months. But it wasn't just the money that made me happy. My newfound independence appealed more to my natural disposition. For the first time in my life I was my own boss, at least on a small scale. I could decide for myself how many people I wanted to approach, how many deals I wanted to seal. And with the money I earned, I was able to make my personal life better than ever before.

Whether you want to call it coincidence or fate—sometimes it is these seemingly small circumstances in life that bring about great change and are the beginning of a whole new way of life. Behind a successful entrepreneur you will often find a long biography of failure and rebuilding, trying and improving. I am certainly not the only one, my biography attests to this rule.

Anyone who narrows success down to a certain event, to a single correct decision, is fooling himself. I could tell you about many other moments in my life that contributed to what I have achieved so far. Whether it was the move from vacuum cleaner sales to the financial sector or the decision to invest in gold, all these choices had a big impact on my professional and personal development.

But successful people must still continue to make decisions. Of course, the first question that had to be answered was: What to do with all the money? Where to put it? How to invest it? I am a huge friend of reinvesting. This not only ensures that the company will continue to develop in the future, it also ensures that the positive effect, the benefit that the company creates for people, continues to exist. This is responsibility, which I take on. I want to improve the lives of my fellow men, I want to be there for people, I want to offer ideas and solutions for their problems and needs. As a twelve-year-old boy I had already begun to feel this responsibility for my fellow human beings. Not just because I had to learn to be independent at an early age due to my mother's alcohol problems. But also because my sister first saw the light of day when I reached the age of twelve. I sensed from that day on that this little person needed protection, security and support. And I definitely wanted to give it to her. Her laughter was so touching that I can hardly describe it. In a way she was also my baby, my little sister. So I took her everywhere with me. On the playground, with friends—I always paid attention to her. My friends were skeptical at first. What was a twelve-year-old doing with a baby? They were more interested in kicking balls around, climbing and playing. But at some point they accepted that my sister was almost always around. I had taken on a responsibility and my friends sensed that this was something special for someone of my age.

To this day, I have a close relationship with my sister. I admire her for her honesty. She is sensitive and empathetic and just like me, she wants to help other people as best she can. My sister has shown me more than any other person that success cannot be reduced to money and prosperity. Of course you can make a lot of people happy, experience great things and enjoy life with a lot of money. But the bottom line is that inner peace, inner contentment is at the core of all happiness.

This really came into focus for me when I was able to be present for the birth of my nephew. After her first failed relationship, my sister met Andreas, who was to be her future husband. Sometime later they married and built a house together. All that was missing was offspring. And he didn't make us wait long. When the contractions began, Andreas said to me: "I don't think I can be there." It was simply too much for him. But I wanted to accompany my sister. That might sound funny. But in this important moment of her life, I wanted to be there. And it was an unbelievable experience. When it was over, I was bursting with happiness. Holding this miracle of nature in my arms, hardly bigger than my hand, filled my entire heart with joy. I was impressed and happy. This child was a real asset for the entire family, a huge success, an enrichment and gift for my sister, her husband and for me.

The financial success of a company can also trigger an enormous emotional wave. Something extraordinary has been achieved. And you are right to be proud of it. But there is a thin line running between pride and arrogance that should not be crossed. Pride comes before the fall. What you want is sustainable success. To phrase it in the language of pop music: you don't want to be a one-hit wonder, but an all-time favorite, a chart-buster and a classic. To keep your feet on the ground, you need a spiritual anchor. A ritual or a place to go that will ground you and show you that life is not always anointed with success. Something that reminds you there are also negative sides to the job.

When I think back on my career, I can also see the tremendous effort and difficulties I encountered along the way to becoming the CEO of Karatbars. That's how it was, for example, when I was involved in the online marketplace project. We operated a website comparable to eBay, but better in decisive ways. Depending on the package the retailer chose, he would have a certain contingent of sales at his disposal, a kind of personal market stall on the net. Instead of a fixed end time for auctions, the deadline for bids was extended by another minute within the last minute. Some hidden extras were also possible. Unfortunately, the project did not produce a return on investment as quickly as planned. And our investors withdrew the support they had previously promised, so that I soon was no longer able to pay the programmer. I would have liked him to remain in the boat and work his way up. But life is not a hit list of your own favorite songs. Without a programmer, no Internet trading platform. That's why we had to close down. And to top it all off, the public prosecutor's office came knocking on the door. The accusation: unfair competition. At first I was shocked. I would never have thought that selling virtual packages could be a criminal offence.

While the prosecutor was turning up the heat, the judge showed more understanding for my situation. He firmly believed that one day I could be successful. He felt my passion, my commitment. I was simply lacking that one, perfect idea that would not bring me into conflict with the law. "You just need the right business model, then you will be successful." I can still hear the sentence given by the judge as if it were yesterday. I managed to get away, in the end, with a bearable fine and the gratitude of getting another chance. With a new, better idea, I wanted to make the big score.

When I look at photos from my youth, I see a young man who tried everything he could, everywhere he could. I started three apprenticeships, I didn't finish any of them, successfully. As a maker of bags, painter and baker—it took me a long time to realize what I wanted. That I was looking for independence and self-determination. I wanted to be my own boss and not just toe the line of my superiors. This self-awareness set the course for my work as an entrepreneur, sharpened my mind and focused my senses on what I really wanted. To create something of my own instead of playing it by the book—that was my destiny.

The memories of my life show me again and again that success is not a matter of luck, but the result of a long journey. Looking back on this long journey puts the success into the right perspective. Memorabilia, photos, friends and family all contribute to a balanced and unadulterated view of one's own career.

It is not uncommon for a completely new circle of friends to appear when success first arrives. Choose them carefully. You don't need anyone at your side who only wants to bask in the glory of your success. True friends do not attach importance to this, because they appreciate you as a person, as a person with all your strengths and weaknesses. Regardless of whether you are currently winning or losing. Compliments are all nice and good. But they can also make you blind and obscure your view of reality. Those who only want to flatter, are not going to help you. You don't need brownnosers around you. Even though the recognition will feel nice. What really counts is sincerity and not being surrounded by "yes" men. Your true friends will not shy away from criticizing you. Maybe even because you are very successful right now. So don't reject them, try to understand the criticism. Listening is what's important here. Good advice is worth more than a hundred words of praise. With your true friends you will find the mental grounding you need.

In addition to the many positive things that success brings with it, there is another challenge that you will face. With success, the expectations also rise. Perhaps not necessarily your own. You may be satisfied with what you have achieved. But from the outside, at least, you will be expected to do more. A successful CEO is, naturally, valued and respected. Success takes him to a new level of perceiving the outside world. One attributes special abilities and talents to him, his successes will be downplayed. And you will be expected to continue to demonstrate great virtues and perhaps even more commitment in the future. It may sound unfair, but with success comes a new, higher standard by which you will be judged. This is why, in times of success, it is more important than ever that you take care in how you conduct yourself. The sums you invest become larger, the risk not necessarily lower. And you will be measured by how you deal with these higher demands. Satisfying these demands is no easy task, but with what you have now learned it is a task you can master.

6.
"Give Something Back to the Society that has Given to You."

At the end of the last chapter, I wrote about the pressures resulting from new expectations that a successful entrepreneur will sooner or later have to face. This is all the more true in today's world, where events and criticism can be immediately disseminated and discussed in the media. You have already gained a few important strategies for dealing with the increased pressure of expectations.

We at Karatbars face up to this expectation by not stopping at what we have already achieved, but by thinking ahead. We look to the future and focus on development potential that still lies undiscovered. Because forward-looking decisions set the course for progress and growth. In 2017, I made the best decision in my entrepreneurial life. A decision that would clearly point to the way forward. The reason was again a crisis, at least one in its beginnings. Karatbars grew steadily until 2017. Our sales have so far left nothing to be desired. But nothing can carry on forever without some kind of momentum. And so I had to face up to the fact that our development was stagnating after a look at the 2017 balance sheets. Progress was not being made any more. New impulses were needed. I was now needed in my role as CEO. I analyzed the current market developments, major trends and technological innovations. The latest trends were in digital currencies, crypto-currencies such as Bitcoin and Ethereum. They allow flexible, fast and above all secure transactions across national borders. They are the future of a globalized financial system. But not all that glitters is gold. Digital currencies are subject to the ups and downs of the markets. Their volatility is a risk. And this was exactly the chance that Karatbars needed. Like in a giant jigsaw puzzle, one idea was fitting exactly into another in my head. We would be the first to create a digital currency that would be fast, freely transferable and at the same time stable. Basing our own cryptocurrency on gold was the key to a new world currency. A project bigger than anything I had ever tackled before. Countries, banks, and people from all over the world will profit from it. The profits we are gaining with this project not only overcame the stagnation in the growth of Karatbars, they also provided us with the financial leeway to add a new dimension that will be able to benefit our charity work. This is an aspect that I would particularly like to highlight in addition to the future development of Karatbars.

As a successful and wealthy entrepreneur, you are expected to be committed to those who are weak and in need in society. Social and humanitarian values play an important role for many people. Quite a few use these standards to shape their lives and decide with whom they do business or what they consume. You only have to look at the hype surrounding organic food production to understand the fundamental changes in the minds of many people. The customer no longer thinks only of his own material well-being, but also considers the social consequences of his behavior, starting with the purchase of certain products. It is no longer just the price that counts. It comes down to a multiplicity of background factors. Would I like to support this company? How and under what conditions was a product manufactured? Can I take personal responsibility for this? These are the questions that a new type of customer, at least in the western hemisphere, is asking himself. With a sincere commitment to social responsibility you can provide the right answers.

But you should not only support social projects because of the public's expectations. If you simply allow yourself to be controlled by your peers, you run the risk of being labelled as weak and influenced by others. As an independent entrepreneur you want to avoid this "puppet image." It damages your reputation and robs you of any authentic charisma.

And authenticity is of great importance especially today in times of shady businessmen, increasing anonymity and the fast pace of life. That's why I'm proud that we at Karatbars have created a concept on our own that combines the old with the new. Gold, 5000 years old, an anchor in times of financial crisis and currency decline and digital currencies, the harbinger of the modern financial world—these are the ingredients to blend for a successful future concept.

Time was running out, the concept was ready. In February 2018, we initiated an Initial Coin Offering (ICO), a process in which people from all over the world invested in our newly created cryptocurrency, thus paving the way for us to go public. After just three months, more than €100,000,000 were raised from the sale of the KBC. We could definitely work with this. That was the starting shot, signaling a new era of Karatbars. The €100,000,000 formed the share capital of our first bank, the first cryptocurrency bank, ever. On July 4, 2018, the time had come. KaratCoin Bank opened its doors in Miami. A momentous date, marking the next milestone in our corporate history. What better date than July 4, the day the United States became independent of the British throne? Independence, freedom, progressive thinking—that is what July 4 stands for. This is at the very core of the US identity. And at the same time the basic convictions that we share with Karatbars. No other opening date could have underscored these ideas better. But that was just the beginning.

We launched our e-wallet app just one month later. Here, customers have the ability to centrally control their financial transactions via Karatbars, whether from home or on the road. Popular digital currencies such as Bitcoin or Ethereum can be quickly and easily exchanged back and forth into the KBC. Every customer, whether a private or business investor, benefits from the low transaction fees. Not to mention the high security standards that we guarantee. Because all of the data stored, is located within a high security server. Hackers don't have a chance.

But that's not all. In order to be able to guarantee a stable investment in the future, we acquired a gold mine together with a partner in Madagascar in August 2018. Our partner now works directly with Karatbars, to ensure even better cooperation in this joint venture and beyond. Gold reserves valuing over €900,000,000 are stored in the mine. The course has been set for a secure future. That this is not an empty promise, as we proved in September 2018. Since then, 135 kilograms have been deposited in our warehouses. This is real value that is underpinned by our payment system and creates trust.

Trust, which is also the cornerstone of sincere social engagement. In order for this to be as effective as possible, it should begin with a phase of thorough customer analysis, starting with yourself. You are not only the company's CEO, you are also a customer in a variety of situations. What is important to you? What urgently needs improvement in the world? Which project is worth supporting and embodies your ideals of a better life?

There's no point in blindly throwing money around the neighborhood hoping that it'll "make a difference." No, with your social engagement you want to show what you and your company stand for, the philosophy you pursue, what values you believe in upholding. Use your social commitment and the time and money you invest in it, to put your attitude into action. You are not just a thinker. You are also a man of action. You are actively involved in shaping and assuming responsibility. That must be the message. These aspects must be at the center of your thoughts.

If the customer can understand why you are involved in a project, he will appreciate the importance of your company's work in a new way. He will, probably, no longer define it as a means to his own personal advantage, but also by the social benefit your company provides. Beyond a personal gain, the business relationship gives him the feeling that he is doing something good. Not only for himself, but also for his fellow human beings. This is what can lift the customer relationship to a new level. Especially, when the competition offers similar products or services, your social commitment can make the difference because your company is appreciated for its contribution to society.

That also rubs off on you as CEO. With the success you have achieved, you have shown that you can look further than the end of your own nose, that you are able to put yourself in someone else's shoes and to think more long-term. This is exactly what is needed when setting up charity projects.

Long-term thinking, looking to the future and achieving our visions—that's exactly what we do at Karatbars. In November 2018 we opened our own, digital exchange where coins can be exchanged. You start the search by specifying how many of what coins you would like to purchase at what price. If the exchange finds a suitable offer, the purchase transaction is carried out. A fast, secure and flexible procedure with KBC as the unit of value. An international exchange where anyone can trade, regardless of location or origin. A unique and unprecedented project based on blockchain technology—the next step towards a globally networked financial system in which everyone can participate. With unlimited boundaries and opportunities. The blockchain is a decentralized database that grows with each transaction. A chain of data blocks that is extended bit by bit. The system is decentralized because each computer connected to the cryptocurrency stores its own one-to-one copy of the blockchain. This makes the system much less vulnerable and, therefore, more secure than centralized databases. This minimizes the risks and lowers the costs of financial transactions. Our cryptocurrency, the KBC, takes advantage of just this feature. This is the future. Investors, ambassadors, and finance ministers have come knocking on our door because they share our vision and because they have recognized the potential of our new innovations. We are laying the groundwork for change in the financial world, deeper and more fundamental than ever before. I will do everything in my power to make this happen.

With the right approach, a win-win situation arises. You can help people in a meaningful way by providing support where it is most urgently needed. Far beyond the economic benefits you are making use of your competencies in the societal arena as well, thus making double use of them.

Remember what I said about success and gratitude? Everyone deserves to be valued, no matter what they do. With your commitment you are expressing this appreciation for your fellow human beings both visibly and effectively. It is a sign of empathy and prudence to give a portion of your money to those who take on important tasks but do not always receive the recognition they actually deserve. Think, for example, of the countless volunteers in charitable organizations. This is exactly where you as CEO can take on a pioneering role. You can show that you are not living in a bubble of statistics and balance sheets, far away from the everyday reality of people lives. Instead, you can "donate" acknowledgement of the extraordinary achievements that make our working lives together possible in the first place. You can become a role model for other company bosses and top managers. Ideally, they will follow suit and a few isolated donations will develop into a broad-based campaign of support. In this way, the world can be improved bit by bit and step by step.

By the way, this is also how trust develops. Who wants to do business with someone who makes you think they are only interested in making more money? The greedy businessman, who unscrupulously wants to exploit every situation he can, without a thought to morality or decency, is more of an enemy than a friend. Negotiating with such people leaves a bad taste in your mouth. One might gain some personal advantage from the business transaction, but the personal discomfort remains. In the long run, this does not lead to a long-term business relationship.

No, the ethics of an honorable businessman are as different and as simple as they are concise. It's all about give-and-take. There are two sides to every coin. Like yin and yang in Chinese philosophy, like victory and defeat in sports, and like joy and sadness in life. These basic distinctions also apply to business life. There is no taking without giving. Anyone who only takes without giving will soon no longer be able to receive. Every business relationship, like every human relationship, is based on reciprocity. You give something because you get something in return and vice versa. You pay attention to your customer and he gives you his trust. You generate income from society and give back part of it to improve the conditions there. This is how you remain a part of the system. Part of a cycle that has supported humanity for centuries and has enabled us to create such a wealth of life and diversity, of ideas and innovations. As a successful manager you are an important engine for driving societal progress. Do not shirk this responsibility, but approach it with a proactive attitude. Create the world by means of the possibilities you now have at your disposal instead of stashing away your resources. Stagnation is a step backwards. The world needs new impulses. And you can give them.

This also holds great potential for your own personal happiness. I know from experience that it is a very simple pleasure giving something back. Seeing how the world changes for the better and feeling that one has contributed to this. There is so much suffering on earth. Whether we're talking in large or small dimensions. Wars, natural disasters and famines dominate newspaper headlines. But there are also many problems hidden in the seemingly small things. The things that are not immediately visible, which don't get widespread attention. I often think back to my childhood and my mother's alcohol problems. I am sure that there are still many children in Germany who find it difficult to get the support and strength they need to prepare themselves adequately for their future lives. It is exactly this that can be achieved in social projects, at least partially. What I mean to say is that there are many ways in which you can make the world a better place through your commitment and through your dedication. You can promote social well-being and live up to your own standards and responsibilities as a successful entrepreneur. In the best case scenario you can ensure that not only financial resources, but also your personal authority flow into your charity projects.

Following on from our Goldcard series, for example, we had a charity Goldcard created at Karatbars. These are unique items with a special value. Instead of the usual look, the charity Goldcards carry special designs from Real Madrid and MotoGP, with whom we had a cooperation. Another card in this series features US actor Vincent De Paul and myself. We are putting our names behind this campaign and would like to contribute to its success. This card also stands for the corporate philosophy of Karatbars. Especially, for the part that is about "giving back." This unique Goldcard will be auctioned off among our sales partners for a good cause. This is how we also convey the idea of taking responsibility and showing solidarity with our fellow human beings to our partners and customers. We take one of our products, modify it a bit, and create some charitable added value in addition to the purely economic one. This is how charity works.

Was everything really better in the old days? I don't think so. The problems were perhaps different. But the crucial difference is that today we can learn about the suffering in the world through the Internet and through social media. That should inspire us to become active ourselves. To stand up for those who are weaker, even if it is sometimes only small gestures of helpfulness here, or a word of encouragement or support with very basic things there, to make our world a bit better. To promote this, we have developed the World of Charity Card. A Goldcard, which is meant to remind the card holder to stand by his fellow men. The helper can pass the card on to others to remind them of their social responsibility. The more people who join in, the more we can achieve. Charity means working together to help those in need.

When an entire company becomes socially active, transparency is the highest priority. The interested customer wants to know concretely and comprehensively which social projects you are supporting. Transparency builds trust. In any case, you should avoid giving the impression that the money you are investing will end up going back into your pocket. These kinds of methods only damage your corporate image in the long run. Lack of credibility is the killer of successful business relationships. Good business transactions and long-term relationships can only be achieved with mutual trust.

It can be very useful, if, instead of simply donating the money, you understand exactly what is happening with it. I don't want to cast doubt on simple financial donations. But I have the impression that many people make it a little too easy for themselves. They give a certain amount of money to an organization, give themselves a clear conscience for a few days and then hope that something useful will happen with the money donated. The scandal surrounding the Oxfam employees accused of sexual abuse in developing countries has shown that this way of thinking is rife with fallacy. Of course, it takes time and energy to become a part of each of the projects you support. But you will only have a clear conscience that your money is going where it is most urgently needed when you can clearly demonstrate this credibly to your customers and business partners. The effort that you invest in charity involvement at the beginning pays off in the form of a corporate image of integrity that you can then build on in the future in good conscience.

That is why we, at Karatbars, have started to become actively involved in the projects we support. We want to see what happens with the money, how it is actually used and what concrete effects we can achieve with it. Taking a closer look is indispensable in financial matters. I have had this confirmed so many times in my experience as a financial advisor. If something really meaningful is to be done with money, a long leash approach is the wrong thing to do. A quickly said "You'll manage" may sound nice, but you will never know whether the capital invested has the greatest possible effect. And it is exactly this effect, this impact, that it is all about. Regardless of whether it is about the expansion of your company or the effective support of a charitable project. You have to make sure that the resources used have the greatest possible effect.

It can also make sense to design and subsequently implement your own projects in partnership with charitable institutions. We are currently planning an awards ceremony together with UNESCO, for particularly emotional personalities worthy of support. The awards will be endowed with prize money and are intended to honor people who have already achieved great things for the good of society. We want to motivate them to continue using their talent for good causes in the future. We want to support people

139

who have shown that with their drive, their will and their talent they are able to change the world for the better. It is precisely such people who deserve support. For they are the thought leaders of a better world, of progress for the good of everyone. We want to bring attention to them with this prize. On the one hand, because they have simply earned it through their extraordinary performance. On the other hand, because they can become a role model for others, possibly new generations, who also want to stand up for other people.

The effectiveness and efficiency of the charitable projects will be evaluated prior to the awarding of the prize. Of course, all social engagement has the goal of improving people's lives. But not every project does justice to this goal. Large organizations have the problem that many donations disappear into the depths of their bureaucracy without the needy receiving what was intended. The path to the actual recipient is far too long and the amount of money is reduced each step of the way. In order for the support to amount to more than a drop in the ocean, and be able to make a real difference, the projects you support must be well-thought-out and intelligently designed. We expect our sponsored partners to know where to intervene in order to achieve the greatest possible benefit for society. Just as a stock exchange official considers exactly what shares he wants to invest in, a charity representative must gain clarity about which charitable projects will achieve maximum social impact with the money spent. Under no circumstances do I want to suggest that you can put the suffering and the neediness of human beings into a simple calculation. Everyone has their problems and a different way of dealing with them. But I also know that my potential to help, even though I am a successful entrepreneur, is limited. Somehow you have to decide who or what the money and commitment will benefit. One's personal biography, experiences and contacts certainly play an important role. But you should also become aware of where you can change people's lives as effectively and sustainably as possible.

At first it may sound paradoxical to view the support of charitable projects through an economic lens. After all, it's about providing positive impulses for the world, regardless of economic

success. Companies are, therefore, often accused of only stag-
ing social commitment in order to polish up their reputation and
thus further increase their turnover. Naturally, this thought may
be behind some charitable activities. But that should not lead to
socially engaged enterprises getting a reputation per se of being
untrustworthy or somehow dubious. Even if there are economic
considerations behind the charitable engagement, this does not
make the societal benefit of the donation or campaign of support any less valuable. If social and economic goals go hand in
hand, the chances of success for long-term support become much
higher. Both sides benefit and are interested in further cooperation. A project with two winners. What could be wrong with that?

It may be morally better to support charitable projects, without economic gain. I agree with that. People's suffering and need
should not be exploited for economic purposes. But at the end of
the day, it is important that these people get the help they need.
Even if it helps a company achieve their economic ambitions, the
social benefits of the support don't go away. It is always better
for charitable engagement to take place. Better, in any case, than
doing nothing at all out of a sense of fear that one could morally
have their back against the wall.

I think most people would agree with me when weighing up
the pros and cons of this issue. Much more detrimental to the
credibility of your company would be contradicting your charitable
donations in your business activities. No one will believe in you if
you are giving money to charitable projects in Third World countries, but at the same time exploiting the low wages and health
and safety standards in those countries. I do not want to condemn
anyone or try to take the moral high ground. Everyone must take
responsibility for their own actions. But if charitable activities and
corporate strategy diverge in such a contradictory way, you will
soon be faced with a lot of critical questions and distrust. Therefore, consider the planning of charitable activities as an integral part of your overall corporate planning. All too often, social
engagement activities are neglected in companies. These things
have nothing to do with the immediate success of the company,

nothing to do with the operative business, they are just a bit of image cultivation. But an opposite way of thinking should find its way into your company. Because the most successful way, for both sides, is social commitment that is embedded in the corporate structure, with coordinated organizational and planning processes and optimum coordination of the projects. Charity should be taken seriously.

I have, therefore, given this task to my sister, a person in whom I have particularly great confidence. I know that these matters are in good hands with her at the helm. I can count on her, fully. And she is capable of leading our charity department, like no other person I know. A willingness to help and a commitment to others has always been close to my sister's heart. It makes her happy to be able to support other people. That was the case with her work as a geriatric nurse and it will be the same with Karatbars. I am convinced of that. I don't know exactly where this incredible passion for helping comes from. But sometimes I think that I also contributed to her great willingness to help when I took care of my sister at an early age, took care of her and protected her as if she were my own child. It was the time when my grandmother and mother were suffering from alcohol problems. I took responsibility for my sister, took her to the playground and possibly even saved her life, when she fell ill with Krupp's cough.

Perhaps at the tender age of one, my sister had already felt what it means to stand up for others, to assist them in the most difficult moments of their lives. Perhaps a need developed deep within her that what I had done for her should be shared with others. Perhaps in those critical moments she felt the value of helping others when they were no longer able to help themselves. I am convinced that her great qualities also stem from those childhood experiences.

What happened between me and my sister in those hours is also conceivable in larger dimensions. This is the mission of good charity work, in my opinion. Creating perspectives for people who are living on the margins. Helping people when they see no way out or are too weak to move forward. My sister shares this vision, not just with her intellect. She feels it with all her heart.

Let me say a few words about how you can get the optimum benefit from charity work for your business. Social project management should not only be embedded in your corporate structure, it should also be accompanied by professional press work. Experience has shown that your biggest critics and those who make the most difference to the credibility of your company are journalists. Every free country in the world has a critical public. And journalists take on this task responsibly by repeatedly poking their fingers into society's wounds. Modern journalism focuses on grievances and scandals. This is how the media world works. But with intelligent press work you can highlight the positive aspects of your commitment and thus convince the public of the value of your company. A positive and serious critique penned by a journalist ensures far more credibility than any newspaper advertisement, no matter how large, for which you yourself are responsible. Because the message of a positive article is: An independent person, trained in critical questioning, has praised your work. While in an advertisement you only praise yourself, journalistic articles are generally regarded as an objective source, as an indicator of the usefulness and sincerity of social engagement. If you have carefully selected your project and the charity campaign fits well to your corporate strategy, you have nothing to hide. On the contrary: Be transparent and show what effect your support has had. Especially, with the right occasion, such as a prize-giving ceremony, the media can come and report on it. In this way you can convey an extremely positive and at the same time authentic image of your company to a wider public.

All of these factors attest to your credibility—as an entrepreneur in a professional environment and as a person in private. And credibility is the foundation par excellence for having a good relationship with present and future customers. It is the foundation on which the trust can unfold that every sustainable business relationship depends upon. This is all the more true in my sector, wealth management. As a rule, people are handing over to you an important part of their financial assets. But trust plays an essential role in all other economic sectors as well.

In the digital world, the threat exists of overlooking this point all too easily. Communication has become simpler, faster and possible over long distances. But this also goes hand in hand with anonymity, a superficiality in the relationships of people who are only in contact via the Internet. I try to counteract this in a targeted way. At big meetings all over the world, people are able to see that real people, with sincere characters, stand behind Karatbars. We have to show our physical presence so that people can see for themselves that we are serious. It's not a rip-off or Internet fraud. It's about responsibility and conviction in one's own company. That's what people sense when you let yourself be seen in public.

And this also includes letting your social engagement be seen. Those who stand up for others stand out. Charitable support services lend depth to your profile as an entrepreneur. You are no longer just a CEO striving for profit maximization and customer well-being, but a man of the world who can see other aspects far removed from the world of numbers and investment returns. It is precisely this dimension that your charitable activity should express. You will not achieve this with a classical model of entrepreneurship.

And be honest, is your only goal in life to make more and more money? Life holds so much in store for those who are open to it. Isn't it much more the awareness of having used one's money wisely, that is of real value to us? And isn't it much more the gratitude of the people supporting us that gives life a new meaning? Why accumulate all this money if you aren't going to put it to good use? In capitalism, money can be translated into deeds, it sets things in motion, it's a driver of change. You can shape these changes. And in such a way that everyone benefits from it.

That is the expectation that you are confronted with and that you should live up to. This is all the more true if your business activities leave a mark on this world. For example, if you are active in the production of goods, you need awareness and appreciation for the natural resources you are taking advantage of. Not all business activities are positive. One should also face up to the critical aspects of one's own business activity and not compulsively negate them. An open approach to the sometimes less than positive consequences for our natural and social environment is the right path towards authenticity and honesty.

We, at Karatbars, are also aware of the sometimes problematic working conditions in gold mining. We take our responsibility as gold traders and, therefore, part of this supply chain, seriously by trying to make improvements. Much can be attributed to the large economic and political power gap between industrialized and developing countries. But this must not be used as an excuse to shift responsibility onto "those up there." Decisive improvements can be made to an individual's life situation, even in the smallest of ways. Politics and business must go hand in hand. This is another reason why we are in contact with politicians from all over the world. It's about finding common solutions that can only be found when we take a holistic approach, when we uncover the connections and causes of grievances and identify starting points. Even though this cooperation constantly presents us with new challenges, far removed from the economy, we remain true to this path. Giving up is not an option. I have learned that as an entrepreneur and that is part of the philosophy of Karatbars.

And I have learned to believe in the great potential of gold. Throughout my life I have been passionate about various business models. From vacuum cleaners and smoke detectors to various types of life insurance, I have sold a lot of different kinds of products. But none of them opens up the possibility of improving the world on such a large scale as gold. As you can see from my previous statements, I am a great advocate of gold. Because I believe that gold can effectively resolve core conflicts. Paper money as a means of payment is the norm all over the world. But it brings with it considerable instabilities. Just take the current currency crisis in Turkey as an example. Prices for the most basic products, such as paper, have almost doubled. Who can guarantee that tomorrow my money will still be worth what it is today? International crises, powerless or unscrupulous politicians are causing inflation rates that clearly diminish the prospects for economic stability. In Central Africa, wars are waged over natural resources, including gold, because they promise wealth in political systems that are permeated by corruption and nepotism. What can be done about these problems?

I'm not saying that gold is the answer to every conflict. The problems are complex and often multifaceted. Comprehensive solutions that go beyond the national arena are needed. In my view, gold is an important building block within any promising solution strategy. With a currency based on gold, financial and also economic continuity could become possible. Unstable economies can be built up, steadily. Resistance to crisis phenomena increases when a currency is based on trust and not on speculation. There is also potential in linking a gold-based currency with the possibilities of modern crypto technologies that we want to exploit with the Karat coin. At this point I would like to mention my close business partner and friend, Josip Heit, who shares the same vision as I do. Together with him, we are expanding blockchain technology along these lines. Because our innovations should serve the people and the state. More stability, more transparency, more trust—all this is what a forward-looking currency must bring with it. This is exactly where Karatbars is at home, this is where our expertise is in demand. So we can do more here than anywhere else, to advance the world. That is exactly the vision I am living for, today.

In this spirit, we are also strengthening our commitment to cooperation with government players. Unfortunately, political and economic instability are still a reality in many countries. The latter, in particular, usually leads to great discontent among the population. It's hard to understand when one's own money, one's own investment, suffers massive losses in value as a result of inflation. And this is exactly what happens on average every 80 years. Every 80 years a currency system renews itself in reaction to the increasing depreciation of the previous currency. People have little reason to be optimistic in this respect. Unless one turns one's gaze from the classic debt currency to other currency models such as that of Karatbars, which is based on gold. We have often pointed out the potential of our model and aroused interest both among government representatives as well as those in business. After all, the basis for economic growth is first and foremost currency stability. Only when people are certain that the money they get today is worth the same tomorrow can trade flourish. But it is precisely this certainty that is lacking in the debt currency model. Particularly, in economically and politically weaker states, there are sometimes landslides in exchange rates that have dramatic consequences for the lives of people in the population. A stable currency linked to the price of gold, as offered by Karatbars, would solve some of these problems. Economic and personal planning security could arise and sustainable development models made possible. Consumption and trade would benefit from this and ensure steady growth.

We see this opportunity and are ready to contribute to the development of a new, stable and mobile currency with our know-how and years of experience in the gold and currency business. This is our moral mission and our professional responsibility.

Nonprofit projects are most promising when they fit to the company. When the company is able to not only give money but also its own expertise. With the expertise concentrated at Karatbars and our financial and human resources with over 600,000 affiliates worldwide, we can initiate positive developments in monetary policy. Whether the impetus is sufficient and actually leads to the successful implementation of our plans also depends on the local political decision-makers. Despite all this, we will never relax until we have achieved our goal.

Irrespective of this, it is important that the desired social engagement is compatible with the company's orientation. By this I mean the professional orientation and its local focus. It seems authentic when, for example, an IT service provider offers support to a local computer course at a neighboring university or high school. The food retailer, who will sometimes have a partial surplus in stock, passes it on to the homeless in that town. A real estate agent could work for the protection of historical monuments in their business area in order to protect cultural treasures and ensure a better quality of life for the people in that neighborhood. These three examples have one thing in common. The entrepreneur is supporting a project that is linked to his work in one way or another. This kind of commitment is impressive. People will notice that you are not just donating for the sake of giving and/or polishing up your corporate image. You show that you really identify with the projects you support by this kind of commitment.

Identification is the next big word. The management, and above all you, as CEO, must be interested in what you want to promote. The general guidelines that I mentioned previously on the topic of successfully forming your company based on your own convictions, apply here as well. You have to be genuinely committed to your engagement to be able to pass it on to others in an emotionally tangible way. Both to your employees and to the wider public. If your social engagement is linked to real emotions, your charity presentation will be much more successful. Conviction is contagious. And it is also much more exciting for you to get involved when you have great passion and feel that you can make a difference.

Once you have found such a project, the receiving party will only really benefit from your support if you are 100 per cent involved. Such partnerships have a much better chance of a long service life. And experience has shown that cooperation gains in quality the longer it carries on. You can achieve more if you can plan long-term and continually build trust. These are insights that you already have from the business world and they apply equally to charity projects.

Let's take a look at the effectiveness of your commitment. Only long term projects allow you to create a better environment that is sustainable. You often hear about aid projects that were over so quickly the amount left over to actually do some good was the proverbial drop in the ocean. It is important to prevent this from happening by finding projects in which you can imagine a long-term commitment and to which you can dedicate yourself with passion and all your heart.

Mere financing alone is not enough to establish a fruitful cooperation. Instead, you should plan joint events with your partners and devise plans on how to best use financial resources. It is about productive cooperation, in which you, as the donor, have a significant role to play.

But you must also give the people working directly on the project enough leeway to develop their own know-how. Trust in the competencies of the other participants, without letting things go their own course. It's a difficult balancing act. The right amount of trust and control must be balanced throughout the cooperation. Extremes should always be avoided.

You can inform the public about concrete projects openly and in a case-by-case manner. This not only gives your company a positive public presence. The project you are supporting will also become better known and more supporters can be found.

In addition to the positive external impact that social engagement brings to your company, charitable work also has a positive impact on your employees, internally.

At Karatbars we have succeeded in building up a steadily growing workforce since 2011. In the meantime, we have 600,000 partners worldwide and the quality of our staff at the core of the company has also increased steadily with our economic success. We have been able to attract some highly specialized financial experts to Karatbars. Employees with specific experience and expertise. Be it in gold trading, the construction of a functioning currency or the integration of trend-setting technologies. It is also thanks to these resources that it has become possible for my initial vision to become a reality, tiny step by tiny step.

With the development of our own gold-based cryptocurrency, we have now successfully entered the digital era. And I am sure that we will play an active role in shaping it. Because the advantages are on our side. I don't know of any other investment model that combines the benefits of gold and digital currency with comparable efficiency. One of the oldest value carriers, gold, is joined together with one of the most modern innovations, the blockchain. The relative stability of the gold investment, supplemented by the flexibility and speed of a cryptocurrency, brings with it enormous synergy effects, which we exploit to the full at Karatbars.

We overcome traditional boundaries by offering stable value with international transactions at much lower fees than the market. And all that in a matter of seconds. At the same time, we guarantee the security of the investment by covering all the assets registered with us with the corresponding gold equivalent. We do this locally by getting involved in the gold extraction process. Karatbars already has a gold mine in Madagascar. This enables us to provide appropriate and humane working conditions on site and to ensure that the assets deposited with us are securely covered by gold. We are no longer dependent on other gold suppliers, but are now holding the reins of our business model in our own hands. To me, borders have always just been challenges in a different guise. We are crossing these borders, and overcoming long outdated barriers with the future-oriented projects at Karatbars. The time of debt currencies has come to an end, the economy is in a permanent process of globalization. We, at Karatbars, are participating in this process by removing obstacles: On the one hand obstacles between countries due to fast and cost-effective transaction possibilities, on the other hand obstacles between individual assets such as gold, money and digital currencies due to our exchange system. In this context, we have also set up a stock exchange that is no longer tied to a national location like the DAX or NASDAQ, but enables international trading in equity securities. All this is new and will have a significant impact on the future of the financial sector. I am sure that this has been made possible by our ideas, our employees and our partners all over the world. By 2020, we want to have a market capital of €500,000,000. That is my goal. And I will do everything in my power to achieve this goal.

In Chapter 3, I talked about the great advantages, indeed the need, of having a well-harmonized team. An important prerequisite for this is continuity in cooperation. That you have the time to adjust to each other, to learn from each other's strengths and weaknesses and to make work processes efficient and productive. You will only have this time if you succeed in retaining high-performing employees. And an effective and authentic charity campaign can make a significant contribution to these efforts. For many people it is an important part of their self-image that their work makes a contribution to the company and society not only economically but also in societal matters. Your employees will be more motivated if they feel that they are taking part in charitable engagements with their work in the company. For this to happen, the company, and above all its management, must act in a socially responsible way as well as acting in a charitable manner. The more your employees identify with your company beyond the purely economic component, the longer they will remain loyal to you. Especially, if the charitable campaigns are local, your employees will quickly feel the positive reactions to your company. In this way, work is transformed from the onerous performance of tasks into a positive activity with social implications, and into a contribution to society as a whole. Especially in these times of individual self-realization, the signal that you send out to your own employees with your social engagement is of great importance.

By now it is becoming clear that charity should be a part of any modern organization. Charity, in the meantime, means far more than just donating a bit of money. Social engagement has become an important part of corporate strategies. It is also a great motivation for CEOs and employees. Social campaigns not only add up to a tremendous advantage for the organizations being supported, they also offer great benefits for you and your company. They engender trust in your customers and business partners. They add a new dimension to your corporate profile and help you express your corporate philosophy through responsible action. Charity creates identification, above all for your employees. It transforms your company into an economically oriented corporation, into a cooperative project that many more people beyond the customers benefit from. If you consider all these points, one thing becomes clear. A well-planned charity activity opens up an unexpected range of new perspectives for the further development of you, personally, and for your company.

Think Big—This means actively shaping and pursuing one's visions and focusing on what needs to be improved. Those who are only "against" something, are, at best, keeping the world imprisoned in its status quo. Our thoughts shape our reality. Things come to us according to the way we think.

In my long career as an entrepreneur, I have learned to focus my spirit on the positive, on the potential to steer progress forward. Those at an anti-war demonstration are focusing on war. Participating in a peace rally would be a much better option. The focus, here, is on what you really want to achieve: peace. Those who internalize this way of thinking can put even the most ambitious goals into practice.

Let us commit ourselves to creating something new, to putting a new idea out into the world and to advancing society. I will continue to do this with Karatbars. The world stands before us like a huge project that we should bring to fruition. Just as I have not yet realized all of my visions, the narrative of this global project will continue to evolve. This future success lies in our hands, alone.

ABOUT THE AUTHOR

Harald Seiz is born in 1963 in Calw, near to Stuttgart. He has been a successful financial and investment advisor since 1978. In 2011 he founded Karatbars International GmbH in Stuttgart and is its managing director. Since then he has consistently and successfully internationalised the business. In 2016 Harald Seiz was awarded senatorial dignity by the Bundesverband für Wirtschaftsförderung und Außenwirtschaft (BWA).

The Future of Money

Harald Seiz

Never before in times of peace has the subject of money evoked the uncertainty it does today. Although, we live in affluence here in Germany, many people begin to ask themselves whether the value of our money is dwindling away. Cash seems permanently under attack as the media bombards us with theories on the 'End of Cash'.

Concerns about the future of money are not without basis: in many countries, massive restrictions on the use of cash have now become a reality, with India at the forefront. Overnight, 86 percent of their rupee reserves were removed from circulation and declared worthless - is cash in the eurozone next?

What is the future of money - a means of exchange, anonymous payment or an opportunity to hoard wealth? How will we pay in the future? What forms will digitization open up to us? And what forms could be forced on us by the state or circumstances, such as a crisis or catastrophe? Are you prepared if ATMs or online banking no longer function?

220 Pages | Hardcover | ISBN 978-3-95972-082-3